OLD-FASHIONED ROSES

150 Favourites

Trevor Griffiths

Trafalgar Square Publishing
NORTH POMFRET, VERMONT

*For all my family and especially my wife, Dixie,
who has been a constant source of encouragement to me*

Cover photograph: 'Poulsen's Crimson'
Cover and text design: Graeme Leather

Printed in New Zealand

First published in the United States of America in 1995
by Trafalgar Square Publishing,
North Pomfret,
Vermont 05053

© Trevor Griffiths 1995
ISBN 1-57076-014-4
Library of Congress Catalog Card Number: 94-60755

All rights reserved. No part of this publication may be reproduced or transmitted
in any form or by any means, electronic or mechanical, including photocopying,
recording, storage in any information retrieval system or otherwise,
without the written permission of the publisher.

Foreword

More years ago than either of us would care to admit to, Trevor Griffiths and I first exchanged catalogues — by today's interpretation an extravagance in terminology for, in truth, they were no more than price lists with the briefest descriptions of the roses. At first glance of Trevor's list it was evident to me, even in those days, that here was not just another commercial grower but a collector and devotee of roses of some significance. The passage of time, however, has proved me quite wrong, for as is now evident, Griffiths' contribution to roses was to be more than just significant, indeed, it has become of major importance both in his native New Zealand and throughout the rest of the world.

He is best known, of course, for his books on old roses wherein, with both his pen and his camera, he has not only given us some wonderful illustrations of the better-known types but some invaluable descriptions and pictures of many of the more obscure, lesser-known varieties. Important though it is to spread the word about roses, such activity could be somewhat sterile if the very variety that whets the appetite of his readers is not available commercially, so it is in the field of conservation that I consider Trevor's work to be most important. In this respect he has preserved for rose lovers many varieties that otherwise would now be extinct. My own catalogue, for example, lists several roses that would not have been available today were it not for him.

Inevitably, I suppose, because of my own special interest, I think of Trevor Griffiths' work as being biased in favour of old roses. However, with him I suspect, as with me, such bias as there is to the roses of yesterday serves only to give us the perspective with which to appreciate the best of those of today, as with this book, putting all roses into context without consideration of age or type.

Since that first exchange of lists all those years ago and the subsequent exchange of varieties, this author and I have become good friends. It therefore gives me great pleasure to have penned these few words in a book which I know, along with his other books, will be one of the most widely referred to in my library.

Peter Beales
Norfolk
England

Introduction

It has been a difficult task to select one hundred and fifty varieties of old roses from over two thousand named types in our collection. First, the roses chosen had to be widely available from specialist growers and, secondly, reasonably good pictures of each had to be available, allowing for the vagaries of weather. There were many that would have been excellent choices but for one reason or another could not qualify.

Many times the question has been asked, 'What is an old rose?' and if one was to be truthful the answer would be, 'Any rose that belongs to the species, Gallica, Damask, Alba, Centifolia and Moss groups.'

The species roses or wild roses are the parents of all roses grown today. While no indigenous rose species are to be found in the Southern Hemisphere, many have found their way there from North America, Europe, China and Japan. Almost all the species roses have generous foliage and single flowers with five petals, and bloom for only a few weeks.

The Gallicas have upright and usually compact growth, unique colours, including striped varieties, which are more numerous than in other groups and, above all, fragrance second to none. Their ability to produce seed that germinates freely, together with an inbuilt hardiness enables them to survive under the most difficult conditions.

Damask roses stand apart from other groups because of their strength of growth, light green foliage and haunting fragrance. They are ideal plants to use between other roses and all types of ornamental shrubs. The early members of the group are of great antiquity and so flower only once each summer.

Albas are distinctive roses with subtle colours, attractive grey-green foliage and an elusive fragrance that has been likened to that of white hyacinths. They are lightly thorned or sometimes thornless and produce white, cream, blush or pink flowers on upright plants. They have been documented from at least 2000 years ago and have been prominent in gardens ever since.

The distinguishing features of Centifolias include deeply serrated handsome foliage; prickles that are usually unequal, whether large or small; delicious fragrance; a pendulous or branching habit; and globular flowers. It has become apparent that there is a great diversity of origins within this type with hybrids from a number of groups including the Chinas and some quite small varieties.

Mosses are very different in that they are the only group to have originated by sporting or mutation and not from seed as in all other groups. They are similar in appearance to the Centifolias except that the green sepals covering the buds are 'mossy'. The Mosses need good growing conditions as some are quite delicate.

These are the genuinely old groups but with the passage of time many roses have been created that have 'genuinely old' characteristics and derive from more recent rose types such as Noisette, Hybrid Musk or Rugosa. To clarify this, the definition of 'old rose' used here is, 'Any rose that has the old form.' To explain this further, it means any rose that has single (five-petalled) flowers to semi-double or very double flowers; they may be flat across the open flower and, with double or very double blooms, may be quartered or muddled in the centre, and can at times have a rolled centre with a green eye.

It must be emphasised that there are tens of thousands of rose species and varieties, and researchers have had the greatest of difficulty with their classification and names. The complexity of the subject may be appreciated when it is remembered that roses were reputedly in the landscape before humans arrived on earth. 'Officinalis', for example, was probably grown for centuries before it gained recognition. Before hand hybridising began, rose seed was sown in fields in Europe. Hundreds of thousands of plants were distributed, with little effort put into the selection of seedlings, and certainly no effort put into the choice of parents.

By the same token, the definition of a modern rose was generally a flower with a high-pointed centre. Of course, there are some exceptions to this. It would also be truthful to say that modern hybridists have not made life easy when it comes to classification, and from the plant's point of view I do not think they care. There is no doubt in my mind that although the early raisers of roses, for example, Descemet or Vibert, did not have the benefit of controlled hybridisation, they definitely had some preconceived programme in their minds of the direction their work should take them.

Similarly, the later breeders of roses also followed definite patterns in their work. Men such as Barbier, Ducher, Guillot, Poulsen, Kordes, Lambert, Pemberton and Bennett all had ideals and in their own ways all followed their heart's desires. Is not this the path of all trail-blazers?

When we come forward in time to the modern era, we cannot but be impressed by the famous breeding establishments of Kordes, Meilland, Swim, Dickson, Poulsen, McGredy, Susuki and many others. Hybridists of all countries worked with Hybrid Teas and later Floribundas but there came a time when these creators of the Queen of the Flowers realised that the further they went, the more strength and vitality they were losing in their new creations, and at different times they looked back to some of the older-known roses for the strong constitution that was now missing. The House of Kordes in Germany was probably one of the first to realise this and do something about it; indeed, a study of their efforts since about 1945 indicates that they may have been doing something all along. Other breeding establishments have followed, either knowingly or unknowingly.

The general pattern has been to introduce an old rose into the programme but, at selection time, when the seedlings put up their first timid blooms, very often those kept for future examination would be of the modern type. Sometimes something rather nice appeared that looked old in form and was different, and this by-product of a planned breeding programme was then introduced.

Nearly ten years ago, my travels took me to many countries, and it was my pleasure to be escorted through the growing fields of many growers and hybridisers. It was apparent even then that there was great change coming in the roses of the future. For the best part of fifty years, Hybrid Teas have held sway over all other types, but with the use of old and species roses in breeding programmes today their future is surely threatened.

David Austin of Wolverhampton in England is one man who has broken the shackles of stereotyping in breeding, and his work is now showing the world the way. Love them or hate them, his English Roses are here to stay, and have made serious inroads into rose production everywhere. His ideal was to use some of the old roses, such as a Gallica, Damask or Alba with some of the stronger modern hybrids, and when the resultant crops flowered, only those with old attributes would be taken forward to the next step in production. His now well-received 'Constance Spry' is evidence of this but, as this superb rose flowered only once in a season, it was not until second-generation plants were created that the ability to flower again was locked into the system. Despite this, 'Constance Spry' has become a popular old-fashioned hybrid with many qualities to recommend it.

A measure of a man's success in life is often assessed by those who copy his work. This is now happening in several countries. Where else can we find roses of consummate beauty, with immaculate old form, a touch of modern colour, the ability to flower over long periods, and powerful fragrances? Surely the best of both worlds.

Yet a few do not perform well under certain circumstances. One remembers some of the unpleasant remarks made about Henry Bennett's roses. But is not this something for the grower to judge, and does not this problem exist with all other introducers of roses? In fact, it is interesting to note that there are many similarities between the life and work of Henry Bennett and David Austin. Both are Englishmen, and both were farmers. General dissatisfaction with farm life and the poor returns from work, together with an inner desire to do something with plants, particularly roses, were the driving forces for both of these men. Each in his own time had the greatest of difficulty getting started in the rose-growing industry, and each was very conscientious and careful in what he did. Problems seemed to dog their very steps, and although there were a hundred years or more between their efforts, they both continued through sheer persistence and determination. Both suffered the indignity of having their work belittled, and both savoured the sweet smell of success in spite of all this. One thing is certain — when the efforts and achievements of these two extremely dedicated rosarians and hybridists are properly assessed and recorded in history, they will both have earned recognition for their pioneering efforts in their chosen fields: Henry Bennett for his inspired work with Hybrid Teas and David Austin for his establishment-shaking work in the creation of his English Roses.

There has at last been a great awakening in the use of roses with old form and colour for interplanting with once-flowering shrubs. Such roses with shrub-like forms have arrived on the scene like a breath of fresh air, and have made tremendous inroads into rose plant production and rose sales around the world. Many gardeners in recent times have taken a long, hard look at their expensive landscapes and have realised that something is missing. About three-quarters of all flowering trees and shrubs have finished their display by mid-summer, and this is where roses with a shrub-like habit come into their own.

Hybrid Musks with their fragrance, clusters of blooms and long-flowering habit, Polyanthas with their clusters of small, double flowers, Teas with their long-blooming ability, fragrance and colour, English Roses with their numerous qualities, and ramblers, climbers, Rugosas and Noisettes all have much to

The striking fruit of Rosa moyesii.

offer. The fact that the older rose types usually grow into a medium to large, rounded shrub with flowers all over the plant is an added attraction to those who have dull gardens from mid-summer onwards. As a result of parentage, many of these roses can also furnish beautifully coloured fruit to brighten up the autumn scene, and in some cases the foliage can colour magnificently. This is true of many of the species, Gallicas and Mosses. Some even have fragrant foliage.

Another factor in the growing popularity of roses with old form and shrub-like habit, including old ramblers and climbers, is their tenacity, durability and performance under the most adverse conditions. This pugnacity or toughness is inbuilt in many of them, being only a few steps removed from the original forms. Many people have discovered this marvellous attribute by one means or another and are now prepared to give away the idiosyncrasies of the modern rose, bush or climber, for the inherent qualities of the older types. The old roses require so little attention that once hooked on them, you are hooked for life. However, as they mostly grow larger than modern roses, it is necessary to allow more space in the garden.

It is my belief that in recent times many people have been frightened away from rose cultivation by those who imply that their care and attention is very complicated. This may well apply to modern roses but it certainly does not apply to the older types. The awakening to old-type roses around the world has been brought about by two main factors: they are easy to grow and easy to look after. Their requirements are basic: deep, well-drained soil with suitable nutrients added to the surface during the growing season, and light pruning, with the removal of dead wood when necessary. The only essential spraying is for aphid control. It is no wonder that the false screen of complexity has now been removed and that at last the truth of ease and simplicity of care has been recognised.

Without getting too involved in lengthy instructions it seems appropriate to offer some advice on rose selection, planting and cultivation to those who may have only recently come to grow roses.

The best method for choosing varieties is to organise a visit to a nursery with a display garden, or to a botanic garden with an extensive rose collection, or to a series of private gardens that have many types and varieties. The flowers must be seen in reality and your choices made from the real thing. After all, you will expect your plantings to last for many years, and a little time taken at the beginning will pay dividends in the long term. The next best way to select your roses is with books, a catalogue or by a visit to a garden centre. However, this method does not give the best information available and often labels or pictures do not show the true colours; they may show the correct form but only for a fleeting moment of time.

Soil conditions are most important for successful roses. Obviously there will be different types of soil in different areas. The important things to remember are a good depth of soil and the best possible drainage. Soils that do not have natural humus should have generous amounts applied to the surface after planting. Sandy and very puggy soils should have liberal amounts worked in well before planting time.

The choice of the actual plants, particularly bare-root plants in the winter, will be out of your hands, especially if purchased by mail-order. But the choice of summer or container plants will be up to you. Medium-sized plants with a good root system will give the best results. They will transplant better and become more easily established under the varying soil and climatic conditions. Large plants with huge stems and soft growth tend to go backwards unless conditions are exceptionally good.

Planting demands several important elements. Old roses will grow in almost any position in the garden. Their old genes allow this, and most if not all of them have their performance enhanced by a greater or lesser amount of shade. Better colour, different and better growth and deeper green foliage can be the result.

The soil should have been prepared some weeks previously by deep digging and removal of all obstacles such as stones, broken bricks or tree roots. If the planting area is in an old garden it may require the addition of some rotted material to improve its moisture-holding capacity. Replacement of soil must be considered if roses have previously been grown in the area. Place the roses in the prepared holes with the union below soil level to give the plant strength; as the plants grow the tops become quite heavy, and in wet and windy weather are blown around and are prone to loss of branches or even breakage leading to loss of the plant.

Roses require feeding and watering during the growing season. Apply plant food or natural or artificial manure around the surface of the soil around the plant. A change of manure from time to time will be appreciated. Watering is necessary for good growth and development of flowers; roses improve tremendously when ample water is given.

Generally speaking, a simple spraying programme should be carried out. Irrespective of the type of rose being grown, some members will require more attention than others. A good fungicide and an effective pesticide should be used — a spray combining both controls may be more practical. Three or four applications during the late spring and summer is all that is necessary.

It is difficult to be specific about pruning and shaping when trying to cover the broad spectrum of all the types. It is necessary to establish a measure of control, but if the plant is doing what you want it to, there is no real reason to do anything. On the other hand, if it is not doing what you want, a gentle reminder with the secateurs can bring it back into line. Careful removal of the old wood that accumulates as the plant ages will keep it replacing branches for many years to come.

Roses are not difficult to grow and maintain even though some people may suggest this for peculiar or pecuniary reasons. Their needs are simple.

The roses listed in this book are but a small proportion of the great number available on the market today. Perhaps others should have been chosen, but those listed here are very good examples of their type. The main qualification was the ability to do well in all situations. After all, this is probably the reason most have survived to the present day.

I have found it practical to deal with the roses in alphabetical order. The name of alternatives is given first, and then the raiser if known. The parentage follows, although sometimes this is not known. Similarly, the date of introduction, i.e., when the rose was first released, is given where possible. The type, height and size are listed next.

Although careful attention has been given to the photographs in this book, one must remember that a photograph only captures the flower for one split second of time. While this usually gives a fair representation of the form of a rose, it sometimes does not give an exact idea of size or colour. As there are so many to choose from, the best advice that can be given is to see the actual roses in the flesh, as it were, before ordering or purchasing them.

List of Roses

'Abraham Darby'
'Agnes'
'Albéric Barbier'
'Albertine'
'Alchemist' syn. 'Alchymist'
'Alister Stella Gray'
'American Pillar'
'Anäis Ségales'
'Auguste Gervais'
'Baby Faurax'
'Ballerina'
'Belle Isis'
'Belle Story'
'Blanc Double de Coubert'
'Blush Noisette'
'Bobbie James'
'Botzaris'
'Bredon'
'Buff Beauty'
'Camaeiux'
'Cameo'
'Cardinal de Richelieu'
'Cécile Brunner'
'Celestial' syn. 'Céleste'
'Chaplin's Pink Climber'
'Charles Austin'
'Charles de Mills'
'Chaucer'
'Claire Jacquier'
'Claire Rose'
'Cornelia'
'Crépuscule'
'Crimson Showers'
'Dainty Bess'
'Desprez à Fleurs Jaunes'
'Devoniensis'
'Dortmund'
'Duchesse d'Angoulême'
'Duchesse de Montebello'
'English Garden'
'Erfurt'
'Evelyn'

'Fair Bianca'
'Fantin-Latour'
'Felicia'
'Félicité Parmentier'
'Ferdinand Pichard'
'Francis Dubreuil'
'François Juranville'
'Frau Dagmar Hastrup'
'Fred Loads'
'Fritz Nobis'
'Frühlingsgold'
'Frülingsmorgen'
'George Arends'
'Gertrude Jekyll'
'Ghislaine de Féligonde'
'Gloire de Dijon'
* 'Golden Wings'
'Graham Thomas'
* 'Great Maiden's Blush' hardy
'Gruss an Aachen'
'Hamburger Phoenix'
'Henri Martin'
'Heritage'
'Hermosa'
'Hugh Dickson'
'Irish Fireflame'
'Ispahan'
'Jacquenetta'
'Jacques Cartier' syn. 'Moreau'
'James Mitchell'
'Jayne Austin'
'Jean Ducher'
'Katharina Zeimet'
'Kathryn Morley'
'Kew Rambler'
'Lady Hillingdon'
'La France'
'La Marque'
'Lavender Lassie'
'La Ville de Bruxelles'
'Leander'
'Leda' syn. 'Painted Damask'

Jardens de Bagatelle hybrid tea "vigorous" NY Bot. Garden "never had to replace"

Othello (old rose)

Zephirene Drouhin (rose pink climbing bourbon)

Eden, a vigorous climber

'Leontine Gervais'
'Mme Alfred Carrière'
'Mme Alice Garnier'
'Mme Hardy'
'Mme Jules Thibaud'
'Magenta'
'Magnifica'
'Maigold'
'Marie Pavie'
'Mary Rose'
'Maxima' — Jacobite Rose
'Mermaid'
'Moonbeam'
'Moonlight'
'Moyesii Geranium'
'New Dawn'
'Nuits de Young' syn. 'Old Black'
'Old Blush China'
'Parkdirektor Riggers'
'Paul Transon'
'Penelope'
'Perdita'
'Perle d'Or'
'Phyllis Bide'
'Pinkie'
'Portland Rose'
'Poulsen's Crimson'
'Pretty Jessica'
'Robin Hood'
Rosa banksiae lutea
Rosa banksiae lutescens
'Rosa Mundi' syn. Rosa gallica versicolor
Rosa x richardii

'Roseraie de l'Hay'
'Rote Max Graf'
'Rugspin'
'Sally Holmes'
'Sanders' White'
'Sea Foam'
'Semi-plena'
'Semperflorens' syn. 'Slater's Crimson China'
'Sissinghurst Castle'
'Sombreuil'
'Souvenir de la Malmaison'
'Sparrieshoop'
'Stanwell Perpetual'
'Sweet Juliet'
'The Fairy'
'The Garland'
'The Pilgrim'
'Tricolore de Flandre'
'Trier'
'Trigintipetala' syn. 'Kazanlik'
'Tuscany Superb'
'Vanity'
'Veilchenblau'
'Viridiflora' — R. chinensis viridiflora
'Wedding Day'
'White Cécile Brunner'
'White New Dawn' syn. 'Weisse New Dawn'
'Wife of Bath'
'William Lobb'
'Winchester Cathedral'
'Windrush'
'Wise Portia'
'Yellow Button'

'Abraham Darby'

Raised by David Austin, United Kingdom
Parentage: 'Yellow Cushion' x 'Aloha'
Introduced 1985
Type: English Rose
Size: 2 metres x 2 metres

It is fitting that such a beautiful rose should be the first. Both parents are comparatively modern, and such is its ability to grow well that some gardeners are treating it as a climber. Its slender, arching growth makes it easy to train either as a shrub by removing the tips of the branches or as a climber or extended shrub by allowing them to develop. The blooms can be quite large — up to 15 centimetres across — and have a delicious fragrance that soon fills the room when placed in a vase. Deeply cupped, they open wide and flat, with lighter yellow on the outside and peachy-pink on the inside, paling with age. The plant is hardy, has a long flowering season, and is healthy at all times.

'Agnes'

Raised by B. and W. Saunders, Canada
Parentage: R. rugosa x R. foetida persiana
Introduced 1922
Type: Rugosa
Size: 2.5 metres x 1.75 metres

The Rugosa family in its original form had only white, pink, cerise and purplish-red flowers before hybridists set to work and intentionally or unintentionally changed things. If you separate the flower from the plant it is difficult to believe that this rose is a Rugosa at all. The blooms are at first deep yellow, paling a little at the intermediate stage, and then becoming creamy. They have a pleasing, unusual scent. Tall in growth and upright, the foliage is small, crinkled and pale green. It is sometimes used for hedges and like some other members of the group does not bear fruit.

'Albéric Barbier'
Raised by Barbier, France
Parentage: R. wichuraiana x 'Shirley Hibberd'
Introduced 1900
Type: Rambler
Height: 6 metres or more

This is a superlative rose in every way. It has all been said before but must with great pleasure be said again. 'Albéric Barbier' originated from one of the best-known hybridists of the time. It is a vigorous grower and can be used for any purpose, including weeping standards. It may appear rather ordinary when first seen, but it is a rose that grows on you when you are able to observe it in your garden. It has a strong, green-apple scent coming first from quite pointed buds, then exquisite flowers of light yellow and, when fully open, creamy-white. It makes an excellent groundcover, and is often used for walls, poles, trellises, archways, covering stumps, and for tumbling over stone walls and banks.

'Albertine'
Raised by Barbier, France
Parentage: R. wichuraiana x 'Mrs A. R. Waddell'
Introduced 1921
Type: Rambler
Height: 6 metres

'Albertine' is probably one of the best-known old roses today. 'Best known' does not necessarily mean 'best of all'. 'Albertine' is seen at its best in United Kingdom and European gardens in mid-summer, where its displays are legendary. Sometimes folk are disappointed because of its once-flowering habit. Nevertheless, it is truly a beautiful rose. A fine fragrance and absolute density of large blooms at mid-season make this a fine acquisition for any garden. The vigorous growth can be trained as a rambler or, by light trimming, it can be shaped as a shrub. The young growth can sometimes suffer from mildew. The reddish-salmon buds open to bronzy, salmon-pink and reasonably double flowers that are about 15 centimetres across.

'Alchemist' syn. 'Alchymist'
Raised by Kordes, Germany
Parentage: 'Golden Glow' x R. rubiginosa *hybrid*
Introduced 1956
Type: Shrub-climber
Height: Up to 4 metres

As this is the first time the classification 'shrub-climber' has been used, perhaps an explanation is necessary. Most of the roses called 'shrubs' or 'modern shrubs' have the ability to remain as a shrub or be encouraged to climb. On my journeys to the United Kingdom and Western Europe, many varieties described as 'shrubs' were actually trained to climb. It seems correct to me, then, to describe them as 'shrub-climbers'. 'Alchemist' has the instinct to either shrub or climb with just a little help. The flowers are of medium size in clusters of three or four, and are golden-yellow at first, becoming orange in the centre as they open. They have a reasonably strong scent. 'Alchemist' flowers in summer only.

'Alister Stella Gray'
Raised by Gray, United Kingdom
Parentage unknown
Introduced 1894
Type: Noisette
Height: Up to 5 metres

'Alister Stella Gray' is one of the mystery roses that appear from time to time with a range of different names. Although it is the cause of confusion at times, this does not detract from its rare beauty. The flowers are small but plentiful, deep yellow in the bud, opening to paler lemon. They are rosette shaped, about 8 centimetres across, and double, quartered with a button eye. They are deliciously fragrant. Years ago this was a very popular gentleman's button-hole rose. The plant is a magnificent sight in full bloom.

'American Pillar'
Raised by Van Fleet, USA
Parentage: (R. wichuraiana x R. setigera) x 'Red Hybrid Perpetual'
Introduced 1902
Type: Climber
Height: Up to 10 metres

Although this old rose does not get good reports from some quarters, the fact remains that it grows in places where no other rose can survive. It has amazing vitality and durability. It can be seen near the sea where from time to time it is showered with salt water; it can be seen at altitudes where the heat of summer and the cold of winter would destroy most other roses; it can be seen in many places in between, and always it looks well with its dark green foliage and its medium-sized single flowers, mostly scarlet with a white eye. It is often found in old cemeteries and goldfield sites — its toughness is beyond reproach. It has no apparent scent and is summer-flowering only.

'Anäis Ségales'
Raised by Vibert, France
Parentage unknown
Introduced 1837
Type: Gallica
Height: 1.5 metres

This rose features many times in New Zealand's early history and it seems to be just as popular today, both in New Zealand and overseas. Of the large group of Gallicas we now grow, 'Anäis Ségales' would be one of the most sought after. The reason for this probably lies in its colour, which is deep mauvish-crimson, paling to a very pretty lilac-pink with age — a very popular and contemporary set of colours for today's fashions. The shape of the bloom is considered perfect, being flat and quite double with a green eye in the centre. It is very fragrant and summer-flowering.

'Auguste Gervais'
Raised by Barbier, France
Parentage: R. wichuraiana x 'Le Progres'
Introduced 1918
Type: Rambler
Height: 6 metres or more

Here we have a delightful rambler that has not been known in the Southern Hemisphere for very long but is more widely available throughout other countries of the world. However, this is one of Barbier's best. The growth is very vigorous, with long canes in all directions if allowed. It is an ideal covering plant. Added to this is the extreme beauty of the flowers, which are about 8 centimetres across and open quite flat. The blooms have short petals and are prolific against the glossy, dark foliage, which always looks well. When open, the flowers are coppery-salmon and lemon, and are strongly scented. What more could one ask?

'Baby Faurax'
Raised by Lille, France
Parentage unknown
Introduced 1924
Type: Polyantha
Height: 60 centimetres

'Baby Faurax' is a lovely colour break in small-growing old roses, and provides excellent colour contrast. The small, very double flowers are arranged in clusters, lavender-purple in colour, pinking a little with age. This little beauty never fails to excite comment from those who see it for the first time. In truth, many of the old dwarf Polyanthas more than hold their own with the more modern miniature and patio roses.

'Ballerina'
Raised by Bentall, United Kingdom
Parentage unknown
Introduced 1937
Type: Hybrid Musk
Height: 1 metre or more

This lovely rose has small flowers that are 3–4 centimetres across, single, rosy-pink on the edge and white towards the centre. They have a slight fragrance. The originator of this rose is given as Bentall who, of course, was the Rev J. H. Pemberton's gardener. It was Pemberton who did much of the original work in the creation of the Hybrid Musks, and when he died much of his breeding material was left unattended. The story about Bentall taking over the nursery is unclear, and it is quite possible that some of the roses credited to Bentall were actually Pemberton's seedlings.

'Belle Isis'

Raised by Parmentier, Belgium
Parentage unknown
Introduced 1845
Type: Gallica
Height: 1 metre

'Belle Isis' is a most beautiful, small-growing rose that is quite distinct. The very double, small blooms open flat and are flesh-pink. They are fragrant with what has been described as a myrrh scent. This attractive, tidy, tough plant is worth a place in every garden. It must have proved a particular attraction to David Austin of Wolverhampton, for this rose and 'Dainty Maid' are the parents of his first very successful introduction, 'Constance Spry'. This was the beginning of his wonderful English Roses. In 1985 it was my pleasure to visit David at his home and nursery. On the wall of a large brick farm shed was a large rose, some 8 metres across and 4 metres high, needless to say in full flower all over. It was 'Constance Spry'.

'Belle Story'

Raised by David Austin, United Kingdom
Parentage unknown
Introduced 1984
Type: English Rose
Height: Almost 2 metres

'Belle Story' is a beautiful rose in a different way. The blooms are large, about 3 centimetres across, open flat, and the petals are incurved with the centre exposing a more than usual number of golden stamens. There are several shades of pink in the flower, and sometimes the blooms have an almost iridescent look about them. The fragrance is strong and pleasant. Several times it has been my pleasure to watch these blooms unfold, and the process is fascinating.

18

'Blanc Double de Coubert'
Raised by Cochet-Cochet, France
Parentage: R. rugosa x 'Sombreuil'
Introduced 1892
Type: Rugosa
Size: 2 metres x 2 metres

This is an extremely beautiful rose of large proportions. The attractive, dark green foliage is always healthy, and sets off the large 10 centimetre blooms, which are silky-white and loosely double. They are deliciously fragrant, and can be followed by large orange fruit, though not usually in great density. This is a classic Rugosa, growing into a magnificent shrub that can be used in many ways. It is an ideal evergreen shrub and makes an excellent untrained hedge.

'Blush Noisette'
Raised by Phillipe Noisette, South Carolina
Parentage: Seedling from 'Champneys' Pink Cluster'
Introduced 1817
Type: Noisette
Height: 4 metres or more

The Noisette group originated with Phillipe Noisette in a South Carolina nursery just after the turn of the nineteenth century, and was brought about by some seedlings sent to his brother, Louis, at his nursery near Paris. The group is quite complicated, including Bourbon, Tea, China and other types with this parentage. This rose is one of the popular ones. It has small, 4 centimetre blooms in large clusters, set in attractive glossy-green foliage. The flowers are rose-pink in the bud, opening to pink and white, and are quite flat with a pleasant scent.

'Bobbie James'

Raised by Sunningdale Nurseries, United Kingdom
Parentage unknown
Introduced 1960
Type: Rambler
Height: 8–10 metres

'Bobbie James' is a chance seedling named after the Hon Robert James. It is thought to be a seedling from *R. multiflora*. From time to time in an established garden seedlings pop up. Some germinate in a group; others grow singly. Most of these gypsies are not of much import, but now and again one has excellent characteristics. This rose is one of those exceptions. I had the pleasure of seeing it growing up and through an old apple tree, over 10 metres high, at Castle Howard in Yorkshire. The festoons of flowers cascading from the tree were unforgettable. This magnificent rose has one huge summer-flowering. The extremely fragrant flowers appear in large clusters and are creamy-white.

'Botzaris' (BELOW LEFT)
Raiser unknown
Parentage unknown
Introduced 1966
Type: Damask
Height: 1 metre or more

Now and then we are told or read about the classical old roses, and mention is often made of roses such as 'Fantin-Latour', 'Maiden's Blush', 'Mme Hardy', and 'Charles de Mills'. 'Botzaris' deserves to be ranked among these old beauties for several reasons. The flower is flat and comes from a fat, pinkish bud, developing into a large (about 10–12 centimetres) quite double, creamy-white bloom that is quartered and sometimes shows a green eye. It has a lovely scent and flowers all over a compact plant with mid-green foliage. It is quite hardy and really is one of the most beautiful old roses. I saw it first in the magnificent garden of the late Valdemar Petersen of Love, Denmark.

'Bredon' (RIGHT)
Raised by David Austin, United Kingdom
Parentage: 'Wife of Bath' x 'Lilian Austin'
Introduced 1984
Type: English Rose
Height: 1.25 metres

David Austin is, and will be, the raiser of some very different and beautiful roses, and 'Bredon' is definitely one of them. In time, its worth will be recognised. It flowers freely over a long period and forms a compact, tidy plant with blooms of an unusual colour: buffy-yellow or perhaps biscuit colour, deeper toward the centre, and paling toward the edges. They are 7–8 centimetres across, quite flat, and almost a perfect rosette formation, made up of many small petals. Densely covered in flowers, the plant is a sight to behold. It may be used as a hedge.

'Buff Beauty'
Raised by Ann Bentall, United Kingdom
Parentage: Probably 'William Allen Richardson'
Introduced 1939
Type: Hybrid Musk
Height: 1.5 metres

For about fifty years this rose has been available to gardeners all over the world but at the same time has been quite a mystery. With the passage of time, fresh information has been made available, and it now appears that the credit for its creation goes to Ann Bentall, wife of the Rev J. H. Pemberton's gardener. It is a superb rose of a beautiful shade of buffy-apricot, deepest in colour at opening and paler later. The attractive, dark foliage looks well with the large blooms of about 8 centimetres across. It has a lovely fragrance.

'Camaeiux'
Raised by Vibert, France
Parentage unknown
Introduced 1830
Type: Gallica
Height: 1 metre

There are quite a number of striped roses within the old types, and the width of the stripes varies. This beautiful rose probably has the widest stripes of all. The plant is reasonably compact and has mid-green foliage. Under good conditions the blooms can be about 10 centimetres across. They come from fat buds and are loosely double. Basically the colour is white, heavily striped with pink and crimson. The colours are at their sharpest when the flower is freshly opened, and later the pink and crimson become lilac and mauve as the blooms age. They are also well scented.

'Cameo'
Raised by de Ruiter, Holland
Parentage: A sport from 'Orleans Rose'
Introduced 1932
Type: Polyantha
Height: 1 metre

The popularity of 'The Fairy' is well known, and producers of rose plants are often amazed at how many are sold each year. 'Cameo' is a member of the same group and is like 'The Fairy' in many ways except for its colour, which is a very pretty fresh salmon. The individual flowers are tidy, rosette-like blooms that appear in clusters and, as they open, weigh the branches down, which is, of course, a characteristic of this family. They are lightly scented and pale a little with age, but if grown in some shade do not lose their colour so much. The foliage is light green on a hardy plant.

'Cardinal de Richelieu'
Raised by Laffay, France
Parentage unknown
Introduced 1840
Type: Gallica
Height: 1.5 metres

This is one of the darkest Gallicas in colour. There has been some doubt about its origins, and also about its classification, but that does not stop this fine rose being well sort after. Apparently this is one old rose that benefits from a good cutting back and the removal of dead wood. The blooms are a deep, dusky, purplish-crimson, quite fragrant, and about 6–8 centimetres across. They are cupped when open and flat across the top, and tend to become a mauvish-purple with age. Whereas most Gallicas have a number of small thorns on slim stems, this rose has few thorns at all.

'Cécile Brunner'
Raised by Pernet-Ducher, France
Parentage: R. multiflora x 'Mme de Tartas'
Introduced 1881
Type: Polyantha
Height: A little over 1 metre

This rose has stood the test of time with great distinction. It has been a magnificent rose from the time of its introduction — over a hundred years ago — and has come to be known and respected. The only rose that surpasses this little beauty is the climbing form of 'Cécile Brunner'. There has been some confusion between this rose and one or two others, partly because nurserymen continually propagate the wrong variety. 'Cécile Brunner' has pale pink, beautifully formed, small, double flowers, produced all over the plant. They are fragrant and could be described as a miniature Hybrid Tea. The plant develops into a compact, tidy specimen, which is probably where it differs from imposters. Three other colours are available.

'Celestial' syn. 'Céleste'

Raiser, parentage and introduction date unknown
Type: Alba
Height: Over 1 metre

In every group of roses there are two or three specimens that are well known and revered around the world. It would not be incorrect to classify this rose as one of them. The blooms are a unique shade of silver-pink; in fact, some have suggested that this rose is the origin of the term 'celestial pink'. They are about 10 centimetres across and are cupped, but quite flat across the top. They are semi-double, and open to show golden stamens in the centre. The lovely shade of pink contrasts well with the greyish foliage, and all this coupled with an upright plant is a joy to see. It has a beautiful scent.

'Chaplin's Pink Climber'

Raised by Chaplin, United Kingdom
Parentage: 'Paul's Scarlet' x 'American Pillar'
Introduced 1928
Type: Climber
Height: 4 metres or more

Although this variety is summer-flowering only, it certainly attracts attention for the duration of its display. The healthy, glossy, deep green foliage sets off the floral effort, and the plant is literally covered with blossoms. Each flower is about 8 centimetres across and semi-double, opening flat to show a good crop of gold stamens. The blooms are a bright cerise-pink, and sometimes there is a white line or two on the centre petals. This is a very tough rose that will grow where most others would not survive.

'Charles Austin'

Raised by David Austin, United Kingdom
Parentage: 'Aloha' x 'Chaucer'
Introduced 1973
Type: English Rose
Height: 2 metres

'Charles Austin' is an early member of the English Rose group. It is still a very fine rose, quite able to hold its own with the more recent introductions. It can be rather tall in its growth, and in many parts of the country is used as a climber to cover a trellis or a wooden fence. The blooms are large — up to 13 centimetres across — quite double, and have a strong, fruity fragrance. They are flat and open in a deep apricot-orange, paling later to a lighter shade sometimes tinged with pink. This rose was named after David's father and, if he was like the rose, he must have been upright, tall, and majestic in his bearing.

'Charles de Mills'

Raiser, parentage and introduction date unknown
Type: Gallica
Height: 2 metres

This is one of the loveliest and most popular of the Gallicas. It is a shame so little is known about it, but at least it has survived for us to enjoy today. When it is growing in nursery rows it is quite distinct, even without its flowers. The young plants are upright with slim stems and light green foliage, and when the large, voluptuous blooms arrive, it does not take long for their weight to force the branches downward. The flowers are all of 10 centimetres or more across and are lightly packed with petals. The blooms are flat and very fragrant and really do look as if they might have been trimmed flat with a pair of scissors. One way of identifying this rose is by moving the petals — in the centre you will find a cavity, which in other roses may be a green eye.

'Chaucer'

Raiser: David Austin, United Kingdom
Parentage: 'Duchesse de Montebello' x 'Constance Spry'
Introduced 1970
Type: English Rose
Height: 1 metre or more

When David Austin released the first of his new English Roses, he named most of the early ones after the Canterbury Tales and, of course, Geoffrey Chaucer had to be in there somewhere. Although the time zones are out of line, I have no doubt that had Chaucer himself been able to see this rose he would have been very proud of it. It has medium-sized, double flowers, bright pink in the middle and paling towards the edge. They are quartered and very fragrant. The plant grows to about 1 metre and can be covered in blooms. When reading about the parents of a particular rose we tend to think, 'So that is where this lovely specimen comes from.' However, in this case, the parents require a little more study. Both roses are summer-flowering only, and yet they have produced the recurrent-flowering variety 'Chaucer'.

'Claire Jacquier'

Raised by Bernaix, France
Parentage unknown
Introduced 1888
Type: Noisette
Height: At least 7–8 metres

This very beautiful, deep yellow rose came to my attention when I first met David Steen, husband of the late Nancy Steen — one of New Zealand's experts on old roses. David wore one in the buttonhole of his suit every day he went to the office, which proves that this variety, grown in the warmer parts of New Zealand, can flower for most of the year. Sometimes it is confused with the yellow form of 'Cécile Brunner'. It has very vigorous growth, excellent foliage, and clusters of small, double, deep yellow flowers that are deliciously scented. It is an excellent climbing rose for many purposes.

'Claire Rose'
Raised by David Austin, United Kingdom
Parentage unknown
Introduced 1970
Type: English Rose
Height: 1.5 metres

In many ways, 'Claire Rose' is reminiscent of 'Souvenir de la Malmaison'. They are a similar colour, although 'Claire Rose' has many more petals. They both tend to spot during wet weather, and both can ball under similar conditions. At first the blooms of 'Claire Rose' have a little orange-apricot in the centre, then they become blush-pink, and then pale to cream. The flowers are large, perfectly formed, very double, and fragrant. The growth is upright with light green foliage. While mentioning the similarity of some English Roses to Bourbons, it seems certain that they will also supersede them.

'Cornelia'
Raised by Pemberton, United Kingdom
Parentage: Unknown, possibly from 'Lessing'
Introduced 1925
Type: Hybrid Musk
Height: 2 metres

It seems that the parentage of 'Cornelia' is not known, but perhaps one day it may present itself. There is no doubt in my mind that 'Lessing', a Hybrid Musk raised by Peter Lambert in 1914, may have had quite a lot to do with it. 'Lessing' comes from 'Trier' x 'Entente Cordiale' and, when examined closely, it has a strong resemblance to 'Cornelia'. This variety is one of a small group of roses of this type that have been very popular, mostly because they have had so much written about them. It is my prediction that many others now being discovered by astute gardeners will become better known in the future. 'Cornelia' has lovely stems covered in a mass of small, double, pink and apricot blooms. The individual blooms are about 4 centimetres across and have a lovely fragrance. The plant can grow quite tall and is suitable as a climber or a hedge. It flowers over a long period.

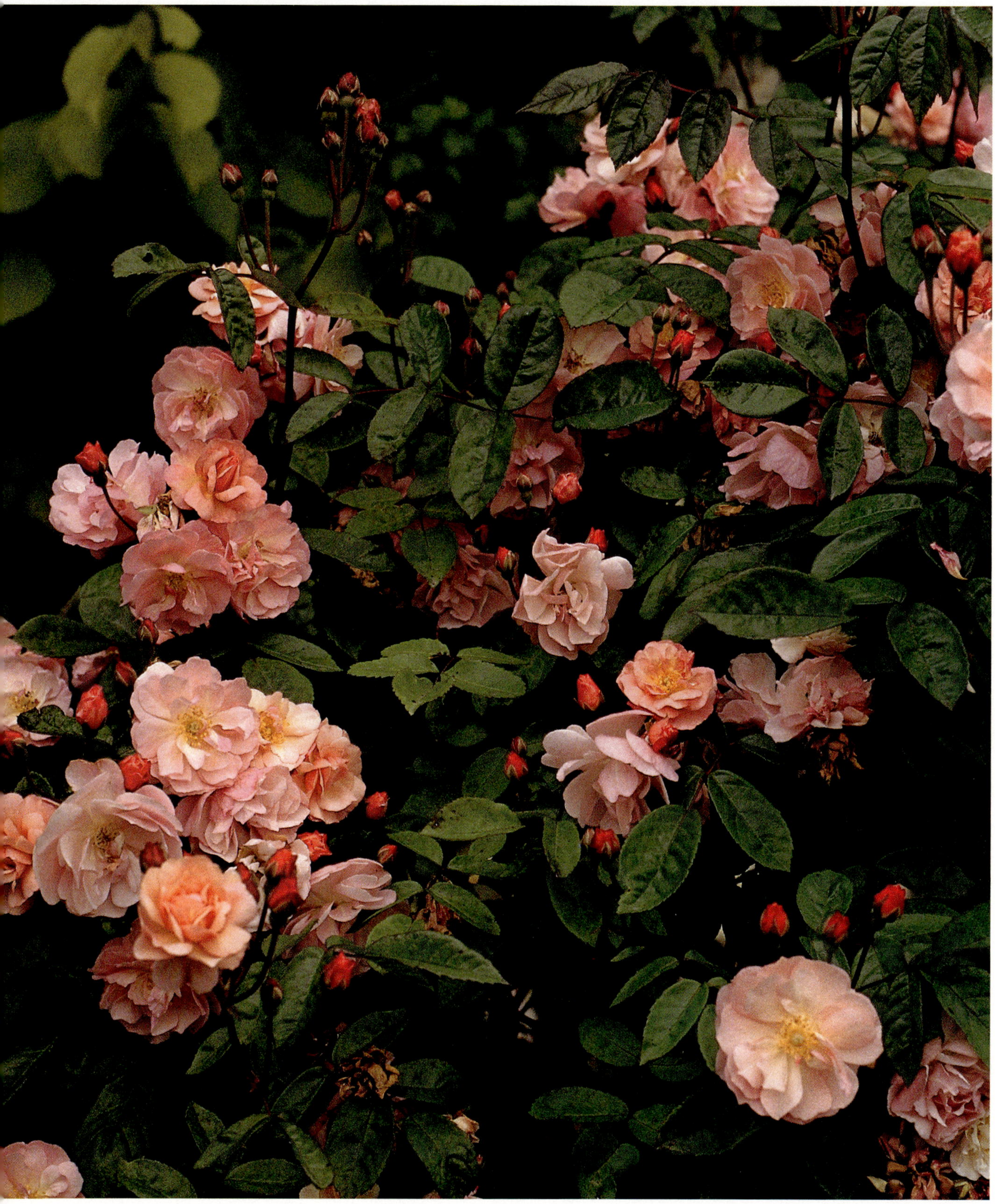

'Crépuscule'
Raised by Dubreuil, France
Parentage unknown
Introduced 1904
Type: Noisette
Height: 4 metres

This is a comparatively unknown rose that has become very popular in New Zealand since it was imported from Denmark some years ago. Although its parentage is obscure, it may have come from 'William Allen Richardson' as at times it bears quite a resemblance. The plant will climb or can be trained as a shrub. In either case, it covers itself with a great density of blossom in a very attractive shade of orange-apricot. The blooms are semi-double, fragrant, deeper in the centre and lighter towards the edges. The light green foliage, deep reddish-brown young shoots and few thorns are some of the attractive features of this popular rose.

'Crimson Shower'
Raised by Norman, United Kingdom
Parentage: Seedling from 'Excelsa'
Introduced 1951
Type: Rambler
Height: Up to 5 metres

The creation of this rose, by intention or otherwise, was a happy coincidence because up to this time other red ramblers, although they gave adequate displays, did not have the ability to flower again later. Each bloom of about 3 centimetres forms a rosette of deep crimson; they appear in large clusters and have a light scent. The flowers appear late in the season but also continue late. All things considered, this is a beautiful and useful rambler

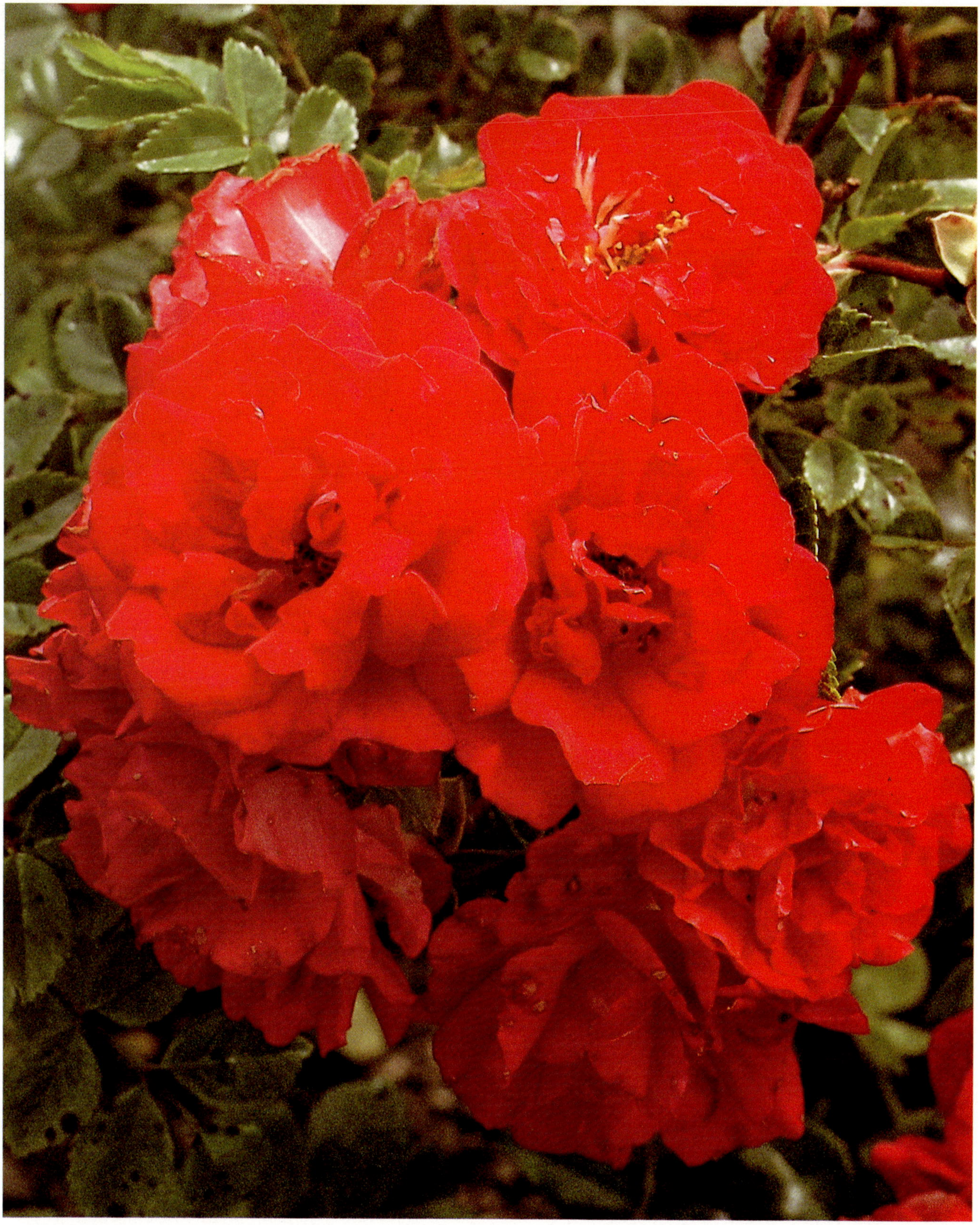

'Dainty Bess'

Raised by Archer, United Kingdom
Parentage: 'Ophelia' x 'Kitchener of Khartoum'
Introduced 1925
Type: Hybrid Tea
Height: 1 metre

Beautiful in every way, this rose is unusual in several respects. It comes from two Hybrid Teas. The blooms are large and single (about 13 centimetres), deeper pink on the outside and silver-pink on the inside. A large boss of crimson stamens contrasts beautifully with the light pink of the flower. Some people take a long time to develop a love for single roses, but when they do they usually collect quite a number. Show them 'Dainty Bess' and they are usually converted immediately. This rose also exists in a lovely climbing form. It has upright growth and a strong, fresh scent.

'Desprez à Fleur Jaune'

Raised by Desprez, France
Parentage: 'Blush Noisette x 'Parks' Yellow Tea-scented China'
Introduced 1830
Type: Noisette
Height: 5 metres or more

Of all the roses that have passed through my hands in almost fifty years of working with them, this particular rose has a fragrance that is different to all others. It is difficult to describe: haunting, delicious, fruity, powerful and all-pervading. This rose also has the ability to remain in flower almost continuously. It will cover a huge area and its pretty, light green foliage sets off the lemon-pink, double blooms that are about 5 centimetres across. At times, shades of yellow and pale apricot appear in the blooms. An unusual feature of this plant's growth is the angular turn of each node. It is an exciting, beautiful and different rose.

'Devoniensis'
Raised by Pavitt, USA
Parentage uncertain
Type: Climbing Tea
Introduced 1858
Height: 4 metres

This lovely old climbing Tea Rose is often referred to as the 'magnolia rose'. Because of its large flowers and, particularly, because of its creamy-white colour, it shows up well against the wall of a dark-stained house or fence. The blooms have an attractive flush of apricot deep down in the centre when they first open. They have a strong tea fragrance and the plant flowers recurrently.

'Dortmund'
Raised by Kordes, Germany
Parentage: Seedling x R. kordesii
Introduced 1955
Type: Shrub-climber
Height: 3–4 metres

The House of Kordes in Germany has produced hundreds of beautiful, hardy roses through four generations of the Kordes family. Without a doubt, every rose in this book could have come from this industrious and successful firm of rose hybridisers. It would be easy to say, 'Why has such and such a rose not been included?' 'Dortmund' is here because of its influence on hybridisers of all countries and, of course, the fact that it appears in the pedigree of many roses. It has deep green, glossy foliage, clean, prickly growth, and overall is very healthy. The almost single flowers are large (12 centimetres), bright scarlet-crimson with white in the throat, and when open the stamens show up well. It has a long flowering period.

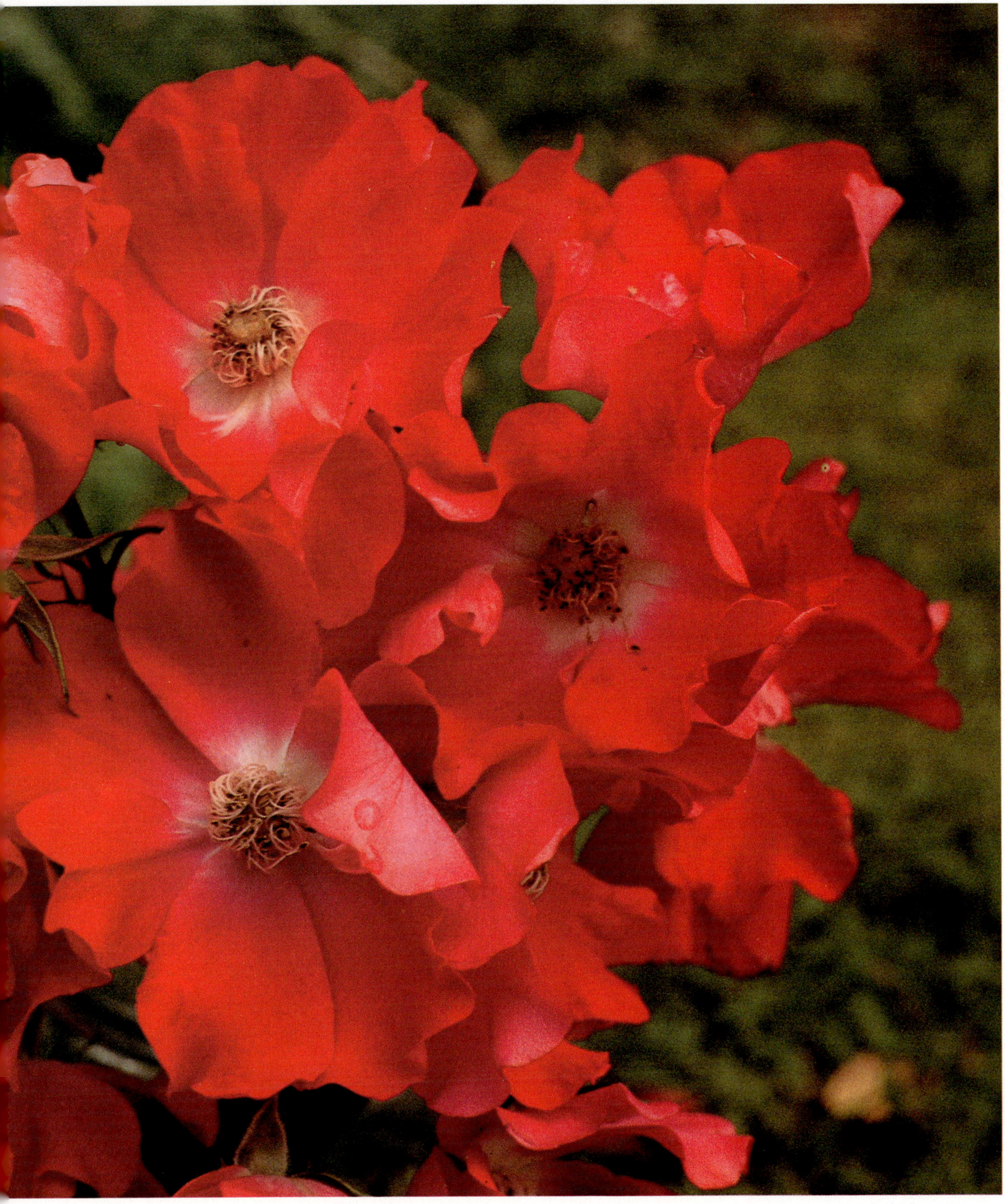

'Duchesse d'Angoulême'
Raised by Vibert, France
Parentage unknown
Introduced 1827
Type: Gallica
Height: 1 metre or a little more

Vibert and his family were close friends of Pierre-Joseph Redouté, and the two families spent a great deal of time together. Redouté was known as 'the man who painted roses', and Vibert as 'a man who created beautiful roses'. 'Duchesse d'Angoulême' was one of them, and was affectionately called 'the wax flower'. It has almost translucent pink flowers that are globular in the bud, cupped on opening, and quite double with a pleasant fragrance. The large blooms tend to hang down on the plant, which is summer-flowering only.

'Duchesse de Montebello'
Raised by Laffay, France
Parentage unknown
Introduced 1829
Type: Gallica
Height: Over 1 metre

Some would consider this variety one of the prettiest of the Gallicas, if not one of the best of the old roses. The shell-pink colour of the flowers is quite exquisite. The fat buds open to cupped blooms that are quartered, quite flat across the top, have a rolled centre, a green eye, and are sweetly fragrant. What more could one ask? Some years ago, on a visit to a contract grower in California, I was surprised to find that the name of his business was The Montebello Rose Co. On being asked if he knew the beautiful Gallica, 'Duchesse de Montebello', the answer was in the negative — another surprise!

'English Garden'
Raised by David Austin, United Kingdom
Parentage: ('Lilian Austin' x *seedling*) x ('Iceberg' x 'Wife of Bath')
Introduced 1986
Type: English Rose
Height: About 1 metre

One often wonders about the naming of a rose and the reasons for the decision. Perhaps, as lovers of roses, we think that we can do better. However, in this case, the name seems to be particularly appropriate, as one can see these lovely blooms melding themselves into an English garden so well that passers-by may not even realise that the plant is a rose. The individual blooms are at least 10 centimetres across, and during the opening process are lovely rosettes, later having a more muddled centre. The colour is at first buffy-yellow, paling with age to creamy-white and lemon at the edges. This is one of the smaller-growing English Roses and has a lovely tea scent.

'Erfurt'
Raised by Kordes, Germany
Parentage: 'Eva' x 'Reveil Dijonnais'
Introduced 1939
Type: Hybrid Musk
Height: Up to 2 metres

The name of this rose always reminds me of my first journey to the State Rosarium, Sangerhausen, in East Germany in 1984. Valiantly, I tried to get the officials to allow me to stay in Erfurt, but they pushed me on to Leipzig. 'Erfurt', the rose, is very beautiful, and gardeners in all countries are just beginning to appreciate it. The foliage is tough and leathery, and the young growth is quite dark and bronzy. The flowers are very fragrant, semi-double, and basically lemon-white with a rosy-carmine edge. Probably 12 centimetres across, they appear in clusters on long, strong stems that could easily be made to climb. There is an almost continuous display.

'Evelyn'

Raised by David Austin, United Kingdom
Parentage: 'Graham Thomas' x 'Tamora'
Introduced 1991
Type: English Rose
Height: Over 1 metre

The House of Austin feels that this rose is one of the very best it has introduced and it is also one of the most fragrant. 'Evelyn' comes from the breeding line that contained the Noisette 'Gloire de Dijon' and has introduced such roses as 'Jayne Austin' and 'Sweet Juliet'. The colour is a little difficult to describe, being neither pink nor apricot nor yellow but a melding of all three colours with a little salmon as well. The blooms are large and cupped and open wide and flat with an old rose arrangement of petals that could be described as a quartering of the outer petals with a quartering of the inner petals. The blooms are extremely fragrant and the plant is free flowering.

'Fair Bianca'

Raised by David Austin, United Kingdom
Parentage unknown
Introduced 1982
Type: English Rose
Height: 1 metre

Everyone who knows the beautiful Damask 'Mme Hardy' will be interested in this excellent addition to the English Roses. At first sight it is similar to 'Mme Hardy', but a closer examination reveals several differences. It does not grow into such a large plant, reaching about 1 metre in height. The flowers are a little creamy at first, becoming white with age. They are cupped, filled with many petals, and quartered, and when fully open have a button eye and green centre. Strongly scented of myrrh, this rose comes very close to the perfection of 'Mme Hardy' but has one major difference — its ability to flower on into the autumn.

'Fantin-Latour'

Raiser, parentage and introduction date unknown
Type: Centifolia
Height: 1.5 metres

I always feel a little sad when discussing a rose such as this. Somewhere, sometime, someone knew everything about this classical old rose, but modern researchers have been unable to discover the origins and story behind this very beautiful variety, loosely classified as a Centifolia. It is unfortunate that the information was not written down. It is suggested that the foliage shows a China influence, although the blooms are more like a typical Centifolia. About 10 centimetres wide, the flowers are pink to blush-pink and are nicely scented.

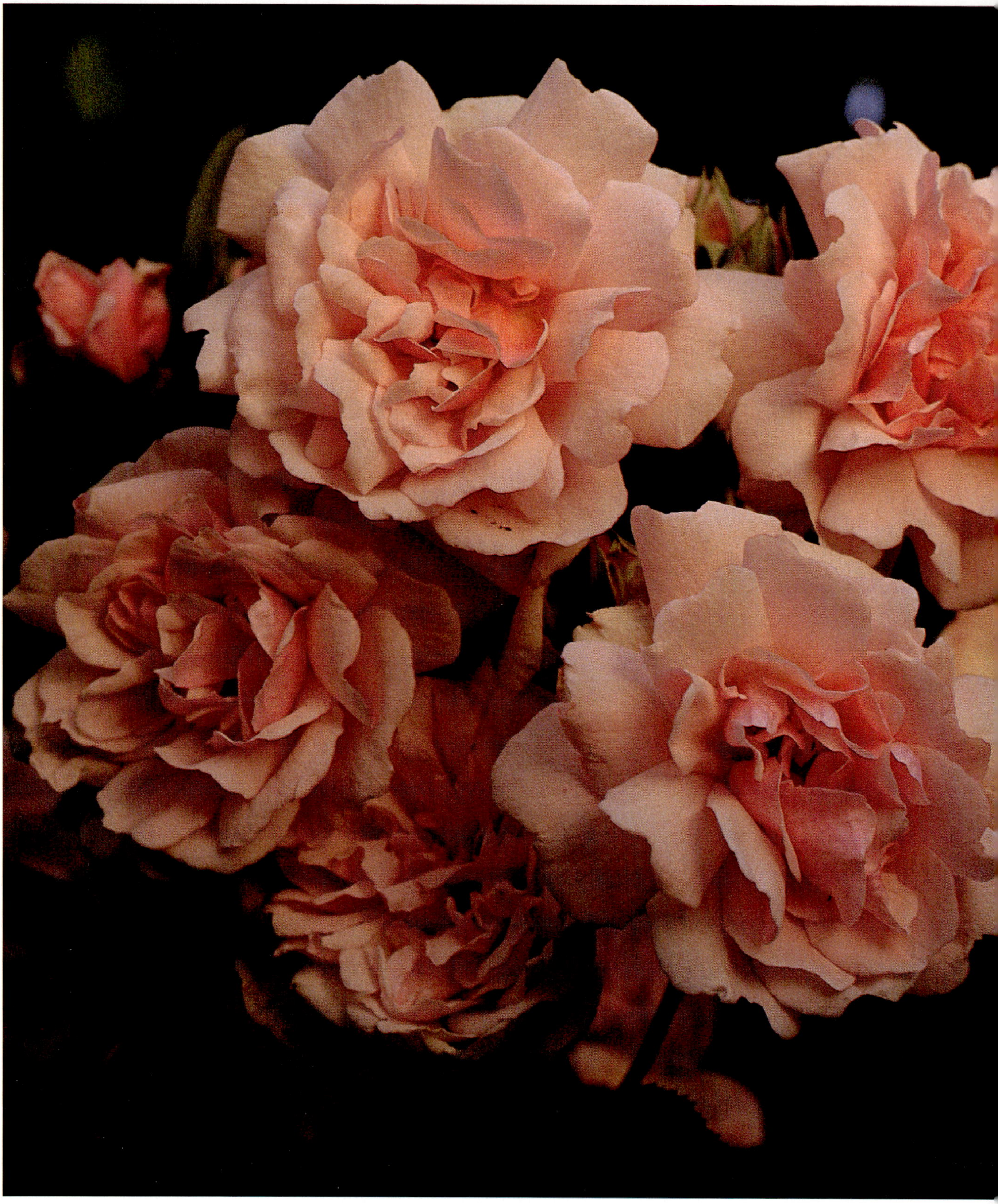

'Felicia'

Raised by Pemberton, United Kingdom
Parentage: 'Trier' x 'Ophelia'
Introduced 1928
Type: Hybrid Musk
Height: 1.5 metres

The Rev J. H. Pemberton has been given credit by some writers for the creation of the Hybrid Musk group. While this may be true, Peter Lambert of Germany worked along similar lines with considerable success before Pemberton. It is significant that this rose came from 'Trier', one of Lambert's roses, and both 'Cornelia' and 'Penelope' have 'Trier' in their ancestry. 'Felicia' always impresses with its medium-sized (8 centimetres) blooms literally covering the healthy, compact plant. They open silver-pink with deeper shades in the centre, and have a strong fragrance and recurrency in performance.

'Félicité Parmentier'

Raiser possibly Parmentier, Belgium
Parentage unknown
Introduced 1834
Type: Alba
Height: Over 1 metre

Once again we have an exquisite rose of which there appears to be no record. If Parmentier of Belgium was responsible for the introduction of the lovely Gallica 'Belle Isis', it seems quite feasible that he or one of his contemporaries was responsible for this rose too. It has rather more greenish foliage than the average Alba, and this may indicate that it is part Gallica. The buds are fat and open with a shade of lemon, soon changing to medium-pink as the flower opens, and then fading to blush-pink and white. They are almost 8 centimetres wide, quartered, flat across the flower and very fragrant.

'Ferdinand Pichard'

Raised by Tanne, France
Introduced 1921
Type: Hybrid Perpetual
Height: Over 1 metre

Quite a number of the old roses are striped, and many are somewhat older than this one. Some years ago, rose purists doubted the authenticity of those roses, preferring to believe that the striping was a result of defects and disease rather than a natural occurrence. Exhaustive tests were carried out by a university in California, which proved that they were healthy, beautiful roses in their own right. This rose is a popular member of the Hybrid Perpetual group. The pink blooms are striped quite clearly with purple and crimson, and are sweetly scented. The plant has vigorous growth and repeat blooming.

'Francis Dubreuil'

Raised by Dubreuil, France
Parentage unknown
Introduced 1894
Type: Tea Rose
Height: 1 metre

This rose first came to my attention on a visit to Sangerhausen in East Germany in 1984. Its colour, scent and form stood out from others near it. The medium-sized blooms are quite blackish in the bud and open first to blackish-crimson, then pale to crimson as they age. They are quite double with a muddled centre and of course are nicely fragrant. The plant has an upright habit of growth and the flowers appear over a long period of time.

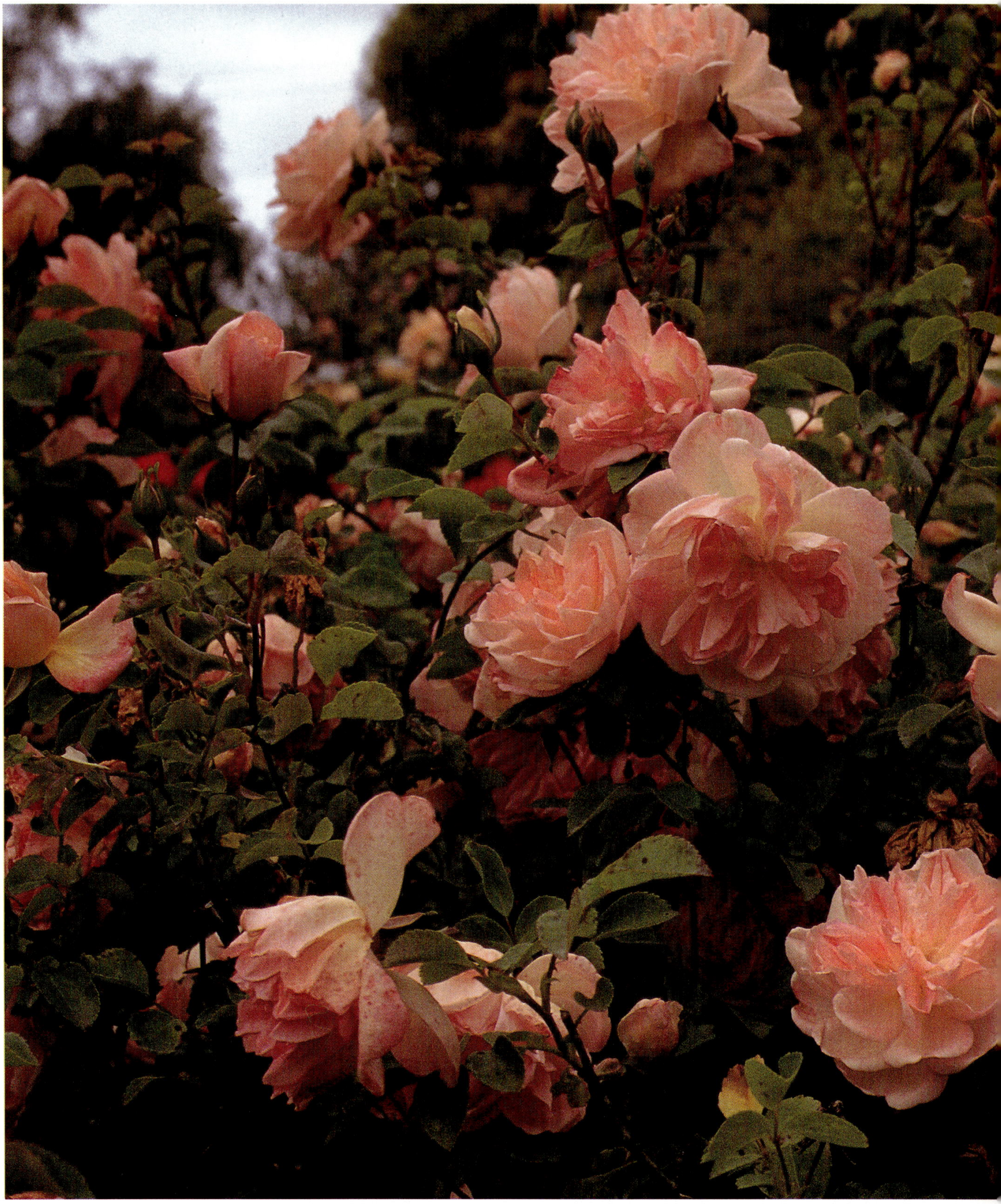

'François Juranville'

Raised by Barbier, France
Parentage: R. wichuraiana x 'Mme Laurette Messimy'
Introduced 1906
Type: Rambler
Height: 8 metres or more

Surprisingly, this rose is sometimes mistaken for 'Albertine'. It has medium-sized blooms about 10 centimetres wide with a strong apple fragrance. They are coral pink on the outside and deeper towards the centre with lemon at the base of the petals. We have this most adaptable rose growing well up into a purple birch tree (*Betula pendula purpurea*) and visitors are quite taken aback by the attractive pink flowers billowing down among the deep purple bronze. We hope that it will grow to the top of this 13–14 metre tree. About the same period of time, Barbier produced three other beautiful ramblers. These are 'François Foucard', which is lemon-yellow in colour and was introduced in 1900, 'François Poisson', introduced in 1907, with sulphur-yellow to white blooms, and 'François Guillot', which has lemon to milk-white blooms and was introduced in 1907.

'Frau Dagmar Hastrup'

Raised by Hastrup, Denmark
Parentage: Rugosa seedling
Introduced 1914
Type: Rugosa
Height: 1 metre

Europeans have shown the world how to use Rugosas, which are natives of Korea and Japan. When travelling by train, especially through countries such as Denmark and Germany, you can see members of the Rugosa group almost everywhere, showing adaptability to natural and man-made environments alike. They are prominent around railway stations and embankments, industrial sites, and in and around motorway turnoffs and median strips. They are tough, and this variety is no exception. The large, single blooms (10 centimetres wide) are rose-pink with lemon-yellow stamens. Both fruit and flowers can be present on the plant at once. This is an excellent rose for hardy situations.

'Fred Loads'

Raised by Holmes, United Kingdom
Parentage: 'Orange Sensation' x 'Dorothy Wheatcroft'
Introduced 1967
Type: Shrub-climber
Height: Up to 3 metres

This fine rose was named after a well-known English gardening expert. We have a plant growing against our office building, and when in flower it never fails to receive favourable comment. The quite large blooms (10 centimetres wide) are almost single and are the clearest bright vermilion shade. They appear in large clusters and flower throughout summer and well into autumn. Added to all this, they are nicely fragrant.

'Fritz Nobis'

Raised by Kordes, Germany
Parentage: 'Joanna Hill' x 'Magnifica'
Introduced 1940
Type: Shrub-climber
Height: Over 2 metres

Although this rose is summer-flowering only, it still merits a place in every garden. It grows into a large, rounded shrub, absolutely covered in flowers when at its peak. In England and Europe it is also grown as a climber. The long, pointed buds are reddish, opening to a creamy-pink with a salmon-pink underside — a beautiful combination of colour — and they are fragrant. The blooms are 9 centimetres across and open flat and wide with stamens showing. A noble rose indeed.

'Frühlingsgold'

Raised by Kordes, Germany
Parentage: 'Joanna Hill' x R. spinosissima hispida
Introduced 1937
Type: Shrub-climber
Height: 3 metres

Much of what has been written about the previous rose applies equally well to this one. There are a number of obvious similarities and even one of the parents is the same. But because the *R. spinosissima* influence has been stronger than that of 'Joanna Hill', we have a tough, durable plant that quite literally will grow anywhere. The flowers are large (about 13 centimetres) and a pretty shade of medium golden-yellow. They are a little more than single and extremely fragrant. Like all members of the Frühlings group they flower early in the spring, when they have a massive display. The growth is upright and tall, the foliage is attractive, and fine prickles cover the stems.

'Frühlingsmorgen'

Raised by Kordes, Germany
Parentage: ('E. G. Hill' x 'Catherine Kordes') x R. spinosissima altaica
Introduced 1942
Type: Shrub-climber
Height: 2 metres or more

There are actually eight members of this Frühlings group, bred by Kordes. Frühlings means 'spring' in German, and all in this group flower early, gaining the ability from the R. *spinosissima* connection. Some writers dismiss the lesser-known ones as not worth growing, but I cannot agree with this conclusion. In their own way and in their own time, they are quite beautiful. This variety does not grow as tall as the previous one and the blooms are large (12 centimetres), fragrant and single. Towards the edges they are rosy-pink and pale to lemon and white in the centre.

'George Arends'

Raised by Hinner, Germany
Parentage: 'Frau Karl Druschki' x 'La France'
Introduced 1910
Type: Hybrid Perpetual
Height: 1.5 metres

With parents like this, 'George Arends' could hardly be anything other than a beautiful rose. The very fragrant, large blooms are medium pink and nicely double. The Hybrid Perpetual group was previously large and at the height of its popularity many hundreds existed, but only a few have survived. 'George Arends' is one of the best.

'Gertrude Jekyll'

Raised by David Austin, United Kingdom
Parentage: 'Wife of Bath' x 'Comte de Chambord'
Introduced 1986
Type: English Rose
Height: Up to 2 metres

A few years ago, when talking about the lady after whom this rose is named, the great rosarian Graham Thomas remarked, 'Jekyll rhymes with treacle'. However you pronounce her name, she was a magnificent landscape gardener. Here we have a magnificent rose, bred from an already successful English Rose and a Portland Rose. It has an extremely good fragrance and has been chosen to be grown commercially to produce the first rose perfume in the United Kingdom for 250 years. When we saw the first bloom of this variety unfold, it was obvious that it was going to be well received. The large flowers are a rich, deep, bright pink, with the petals spiralling as they open.

'Ghislaine de Féligonde'

Raised by Turbat, France
Parentage: 'Goldfinch' x unknown
Introduced 1916
Type: Hybrid Musk
Height: 3 metres or more

The other parent for this rose is believed to be a Hybrid Musk, possibly 'Trier'. We had the pleasure of introducing this rose to New Zealand. It came from my late friend, Valdemar Petersen, of Love, Denmark, who had one of the best collections of old roses in Europe, and, more importantly, his roses were accurately named. This rose almost sneaks up on you. The flowers arrive in clusters of different pastel shades: lemon, yellow, pink, orange, salmon and sometimes red shades. The blooms are small and lightly scented and are produced abundantly. The foliage is light green, and the growth is tall; the plant can be treated as a shrub or a climber. It is sometimes reported to be the best rose in the garden. For something different try this beauty.

'Gloire de Dijon'

Raised by Jacotot, France
Parentage: 'Desprez à Fleur Jaune' x 'Souvenir de la Malmaison'
Introduced 1853
Type: Noisette
Height: 4 metres or more

The parentage is uncertain, but that does not alter the beauty and the durability of this rose. In some places it has attracted almost saintly reverence. There is no doubt it is a treasure, having retained its popularity for nearly 150 years. This is a happy rose in that it begins flowering early and tends to continue blooming right throughout the season. The blooms are about 10 centimetres wide, quite double, at times quartered, and buffy-orange in colour, sometimes deeper, sometimes lighter. The growth is fairly vigorous and the foliage is quite dense when the plant is young.

'Golden Wings'

Raised by Shepherd, USA
Parentage: 'Soeur Thérèse' x (R. spinosissima altaica x 'Ormiston Roy')
Introduced 1956
Type: Shrub-climber
Height: Up to 3 metres

'Golden Wings' is a good example of a rose that exists only in shrub form but has the hidden talent to climb if encouraged to do so. There are examples of this rose being trained to climb to at least 5 metres. Once again, we see toughness coming forward from one side of the parents. The large, nearly 15 centimetre flowers are just a little more than single and are lightly scented. The plant is wide growing with light green foliage, and flowers quite well from mid-season to late. It is an excellent variety for many purposes but does need plenty of space.

'Graham Thomas'

Raised by David Austin, United Kingdom
Parentage: 'Charles Austin' x ('Iceberg' x *English Rose*)
Introduced 1983
Type: English Rose
Height: Up to 2 metres

Once in a lifetime a rose of this calibre is created by a hybridist. The colour is superb: a deep buttery-yellow on opening, paling a little with age. The buds are fat and the flower opens cupped at first and flat across the top. The blooms are large and quite double with a muddled centre. The petals reflex a little with age. The flowers are borne singly, mostly in clusters, and a strong tea scent pervades the whole plant. We know of one plant that had 400 flowers on it at one time. The growth is upright and vigorous with light green, healthy foliage. This is a truly magnificent rose in every way.

'Great Maiden's Blush'

Raiser and parentage unknown
Introduced before 1738
Type: Alba
Height: Up to 2 metres

Whenever the genuine old roses are discussed it is not long before this beautiful old-timer enters into the conversation. It has two forms: the one illustrated has larger flowers (about 10 centimetres) and the smaller form has flowers about 6 centimetres across. They are otherwise similar except for the size of the plant. The flowers are pink to blush-pink, cupped and reasonably double with a pleasing fragrance. Like all members of this group, the plants are very hardy. One of the attractions of this rose is the way in which the lovely pink blooms are set off against the grey-green foliage.

'Gruss an Aachen'

Raised by Geduldig
Parentage: 'Frau Karl Druschki' x 'Franz Dugen'
Introduced 1909
Type: Shrub
Height: 1 metre

This rose arrived on the scene probably before its time. The Floribunda class was not known. It was distinguished from Teas and Hybrid Teas and for many years stood alone, much loved but isolated. When the Floribundas finally arrived after World War II, it had something in common with them, but it was not until the English Roses made their appearance that it at last found some cousins and a family to belong to. It is a beautiful little rose. The blooms open blush-pink and fade to creamy-white, sometimes with the barest trace of pale pink. They are lightly scented, about 8 centimetres wide, and appear from summer to autumn. We also have a pure white form of this rose.

'Hamburger Phoenix'

Raised by Kordes, Germany
Parentage: R. kordesii x *seedling*
Introduced 1954
Type: Shrub-climber
Height: Over 4 metres

The semi-double, large (10 centimetre) flowers of 'Hamburger Phoenix' open wide and appear in small clusters on long, strong branches. The colour is bright scarlet and the blooms are lightly fragrant. This plant makes an excellent free-standing shrub or a low climber, depending on the way it is trained. It was introduced into New Zealand many years ago, but the gardeners of the day were not ready for it. Now, with a new-found love for older-type roses, it is gaining popularity. Ever blooming, it makes a great splash of colour.

'Henri Martin'

Raised by Laffay, France
Parentage unknown
Introduced 1863
Type: Moss
Height: At best 2 metres

'Henri Martin' is probably the nearest to real crimson in the whole Moss family. One could not be blamed for thinking this rose is more modern than its introduction date. The flowers are nicely rounded, reasonably double, and of a medium size. They open an intense crimson and, a day or two later, fade to deep rosy-pink. The blooms are so consistently produced that at times the rounded plant could be mistaken for a camellia. The buds are lightly mossed and the blooms are quite fragrant.

'Heritage'

Raised by David Austin, United Kingdom
Parentage: Unnamed seedling x ('Wife of Bath' x 'Iceberg')
Introduced 1984
Type: English Rose
Height: 2 metres

'Heritage' is one of the earliest of David Austin's introductions and one of the very best. This beautiful variety is as near to a classic old rose as you could get. The rounded blooms are cupped and flat across the open flower. The petals are nicely quartered and the colour is blush-pink. The plant growth is strong, with wiry stems and few thorns. The foliage is deep green and healthy. Under certain circumstances, rust can be a problem in New Zealand, but modern fungicides solve the problem quite easily.

'Hermosa'

Raised by Marcheseau, France
Parentage obscure
Introduced 1840
Type: China Rose hybrid
Height: 1 metre

This delightful, tidy little plant always seems to be in flower. It is thought that there is some Bourbon in this rose. It is certainly a little beauty with smallish pink and lilac blooms that are produced prolifically. They are cupped and have a light scent. There is also a climbing form of this rose called 'Setina', which is very beautiful in its own right, with arching branches covered with blooms.

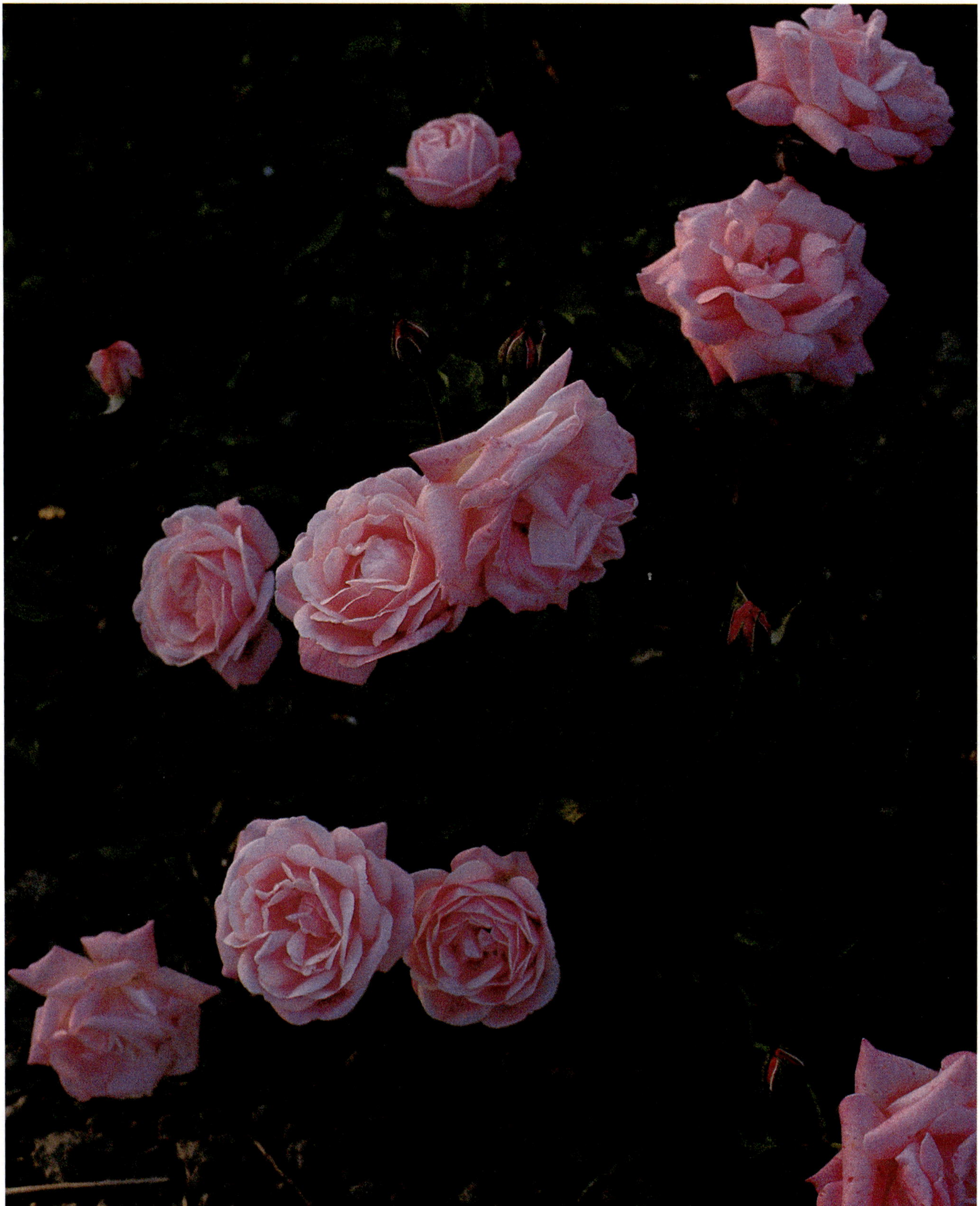

'Hugh Dickson'

Raised by Dickson, Northern Ireland
Parentage: 'Lord Bacon' x 'Gruss an Teplitz'
Introduced 1905
Type: Hybrid Perpetual
Height: Up to 2 metres

Nearly ninety years after its introduction, this rose still commands attention. Not only have many of the old roses survived, but with man's assistance in repropagating, their beauty of form has been preserved for us to admire today. The very large, double flowers are freely produced on a tall-growing plant. They are a rich crimson and very fragrant.

'Irish Fireflame'

Raised by Dickson, Northern Ireland
Parentage unknown
Introduced 1914
Type: Hybrid Tea
Height: Up to 2 metres

Some years ago, while travelling through the suburbs of the city of Christchurch on a balmy summer's evening, we noticed a fine flowering shrub, absolutely covered in flower. On closer examination we found it to be a very large plant of 'Irish Fireflame' totally covered in bloom. It was at least 2 metres high and almost as wide, standing in grounds of an almost derelict house. The blooms of this variety are large (about 13–14 centimetres), single, and very fragrant, with long, reddish-pink buds opening to an exquisite flower of orange to old-gold shades.

'Ispahan'

Raiser and parentage unknown
Introduced before 1832
Type: Damask
Height: 2 metres

When we think of the Damasks as a whole our thoughts perhaps roam to the creation of rose waters and some of the tales from the Middle East that have been handed down to us. If you had room in your garden for one Damask only, then perhaps this rose would be your choice. It is one of the first of the Damasks to open, and one of the last to finish flowering — the period of bloom extends over six to eight weeks. Sumptuous flowers of a beautiful medium-pink quite literally cover the large plant and are strongly fragrant. From a distance this magnificent plant could be mistaken for a camellia in full bloom.

'Jacquenetta'

Raised by David Austin, United Kingdom
Parentage unknown
Type: English Rose
Height: 1.5 metres

This rose has now been dropped by Austin Roses because it really does not conform to the criteria of an English Rose as set up by the parent company. However, many have come to love its simple charm. The flowers are large (up to 12 centimetres) and semi-double, pale apricot in the bud opening to a paler apricot-blush. They are nicely fragrant and appear in handsome clusters.

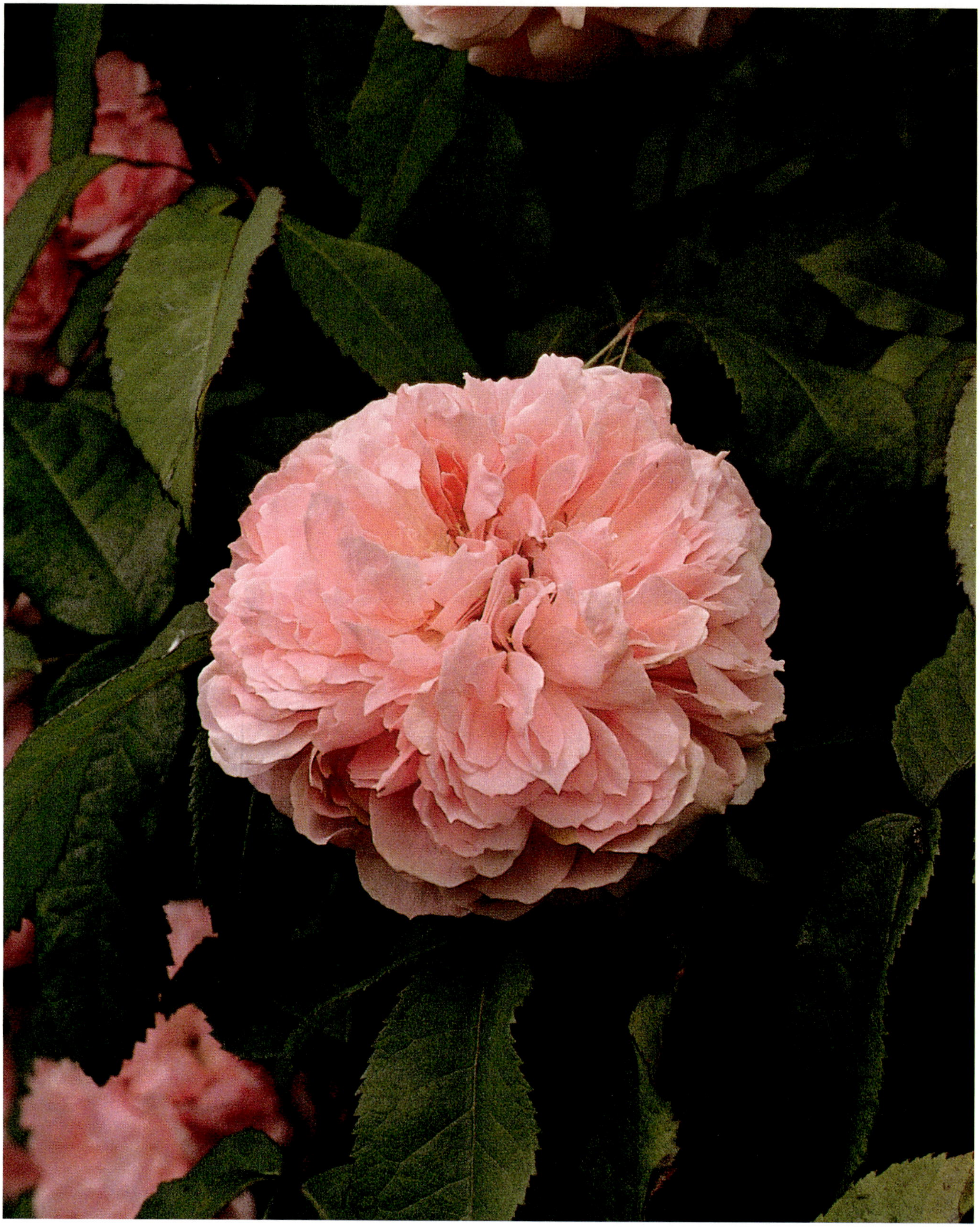

'Jacques Cartier' syn. 'Moreau'
Raised by Robert, France
Parentage unknown
Introduced 1868
Type: Portland
Height: 1.5 metres

The Portland group came from diverse backgrounds and, because of circumstances at the time, they did not last long in popularity. 'Jacques Cartier' is a fine example of the type. A distinctive feature of the Portlands is the way in which the flowers are supported by quite short stems sitting rather neatly in the foliage. The blooms are medium sized, very double and quartered with a button eye. They are light pink, deeper in the middle, and very fragrant.

'James Mitchell'
Raised by Verdier, France
Parentage unknown
Introduced 1861
Type: Moss
Height: 2 metres

This is one of the first Mosses to flower. When it reaches the peak of its blooming season the lovely, large rounded plant is covered with many perfectly shaped flowers of bright medium-pink. The dainty, well-shaped buds are mossed brown-green, and develop into bright cerise-pink blooms that pale to pinkish-lilac. They are medium sized (6–7 centimetres), opening flat with a button eye.

'Jayne Austin'
Raised by David Austin, United Kingdom
Parentage: 'Graham Thomas' x 'Tamora'
Introduced 1990
Type: English Rose

'Jayne Austin' is one of the latest to come off the English Rose production line. As with most establishments engaged in improving their product, this rose signals just that. It has a strong Tea Rose fragrance, and once it breaks into flower it has the breeding to repeat almost continuously. Looking at a picture of this rose it is easy to imagine that it is one of the classic old roses, except for its colour — apricot-yellow becoming apricot in the centre and paling towards the edges. The blooms are medium sized (about 8 centimetres wide), cupped and flat. The petals have a silky sheen and the plant is healthy with pale green leaves. The colour of the blooms is very popular today and this rose will remain with us for a long time to come.

'Jean Ducher'
Raised by Ducher, France
Parentage unknown
Introduced 1873
Type: Tea Rose
Height: 1.5 metres

It is perhaps strange that this beautiful Tea Rose does not appear to have surfaced in the United Kingdom or Europe, and yet it has been recognised in New Zealand for almost a century. It grows over a metre high and just as wide. The foliage is leathery and healthy and the flowers are large and a little more than semi-double. The buds are at first globular, opening wide and ageing from a soft salmon and lemon to a peachy shade. They are pleasantly fragrant with a tea scent. As this is a Tea Rose it may not be very hardy in cold climates.

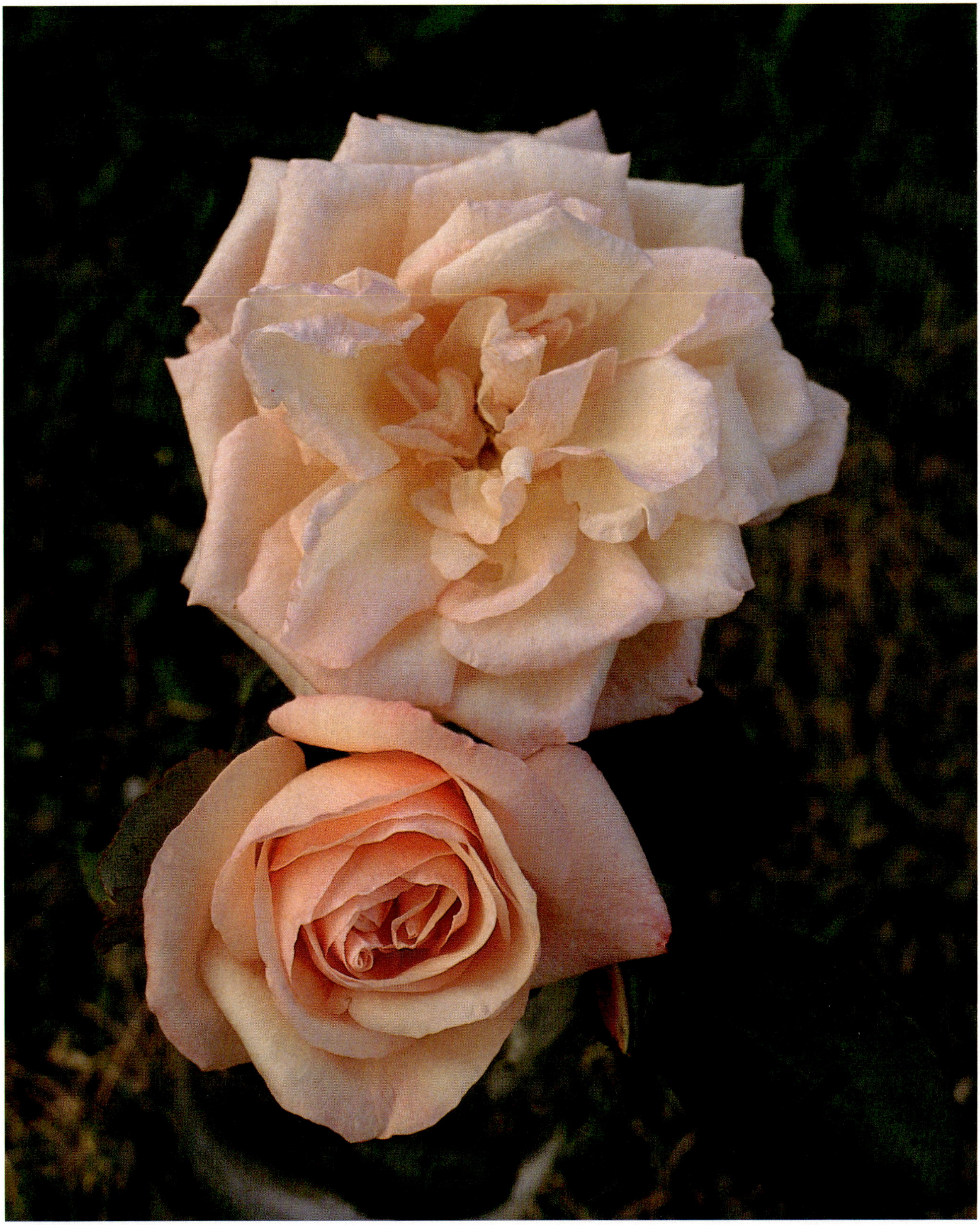

'Katharina Zeimet'

Raised by Lambert, Germany
Parentage: 'Étoile de Mai' x 'Marie Pavie'
Introduced 1901
Type: Polyantha
Height: Up to 1 metre

Now and again a rose shows its pretty little face when many people have thought it lost. This little gem surfaced in East Germany in 1988 at Sangerhausen. Since its arrival in New Zealand it has become very popular. Of compact growth, it will reach a rounded plant of about 1 metre. The attractive, mid-green, shiny foliage and wiry growth supports a charming display of near pure white rosettes, which at first can be quite greenish. The small blooms are packed on a large head and have a pleasant, light scent.

'Kathryn Morley'

Raised by David Austin, United Kingdom
Parentage: 'Mary Rose' x 'Chaucer'
Introduced 1990
Type: English Rose
Height: About 1 metre

Clear, pink flowers, paler at the edges, sit upright on a healthy plant. This is one of the latest of the English Roses and inevitably it will become popular. The flowers are fragrant, large and many petalled, and repeat very well throughout the season. It is named after Mr and Mrs Eric Morley's daughter, who passed away at only seventeen years of age. The Variety Club of Great Britain auctioned the name in aid of the Shaftesbury Homes. It raised thirteen thousand pounds.

'Kew Rambler'

Raised at Kew
Parentage: R. soulieana x 'Hiawatha'
Introduction date unknown
Type: Rambler
Height: 5 metres or more

If you are looking for a rambler that is different from all the others, this could be the one. The foliage is greyish-green, which comes from *R. soulieana*. The flowers are single, bright pink towards the edge, and white in the centre, most likely coming from 'Hiawatha'. The blooms are 4 centimetres across, very fragrant, and appear in prolific clusters all over the plant. It is summer-flowering only, and has its peak on or about the longest day. Different, beautiful, fragrant and vigorous — what more is needed?

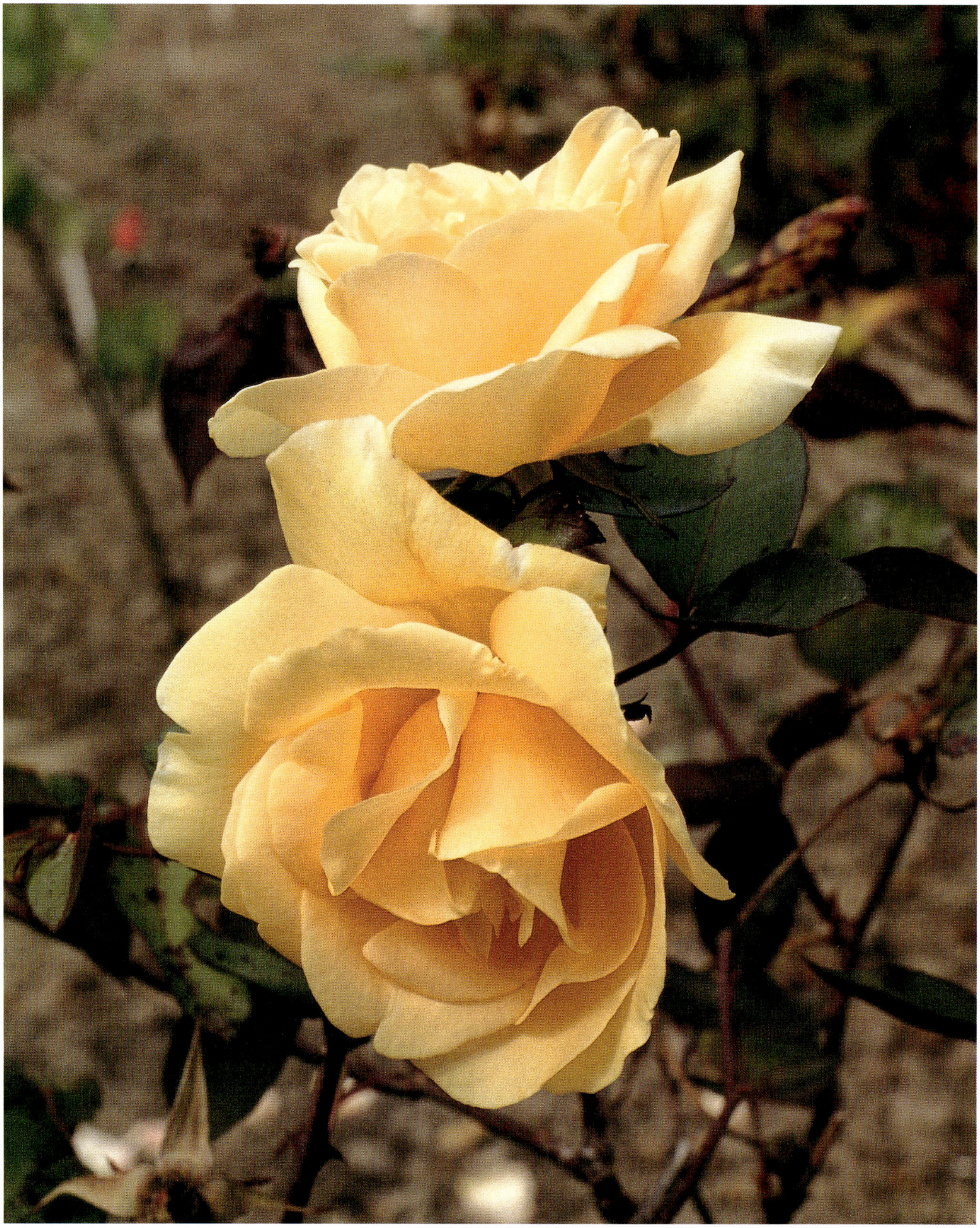

'Lady Hillingdon'
Raised by Lowe and Shawyer, United Kingdom
Parentage: 'Papa Gontier' x 'Mme Hoste'
Introduced 1910
Type: Tea Rose
Height: 1 metre or more

Colour is very difficult to repeat or copy in roses. Even with the many thousands of roses that have been introduced in the last 200 years it is unlikely that any two of them are really alike. Probably this is one of the main reasons why they are so hard to identify. 'Lady Hillingdon' blooms are long, pointed and an attractive shade of warm apricot with a strong tea fragrance. The growth is wiry and tough, reminding the viewer of one of its parents, 'Papa Gontier'. Perhaps the flowers tend to hang down a little when they are fully open but this was true of many of the early Teas. There is also a very fine climbing sport of this rose. In our town there used to be a beautiful specimen, which reached all of 6 metres wide and over 2 metres tall and literally covered its netting support. This wonder to behold was a traffic stopper, but beauty was not in the eye of the recent owner and it is now gone.

'La France'
Raised by Guillot Fils, France
Parentage: Possibly 'Mme Victor Verdier' x 'Mme Bravy'
Introduced 1867
Type: Hybrid Tea
Height: About 1 metre

It is always possible to express doubt about the authenticity of the names of roses. In fact, this seems to happen quite frequently. This rose has had some doubt expressed about it, but we must retain the present name until such time as proper credentials are produced. 'La France' is nevertheless a very important rose. Reputed to be the first Hybrid Tea, it was probably many years before its time. A high, pointed bud develops into a magnificent flower of pink and silver that is nicely fragrant. Although not so eyecatching today, it remains historically important.

'La Marque'

Raised by Marechal, France
Parentage: 'Blush Noisette' x 'Parks' Yellow Tea-scented China'
Introduced 1830
Type: Noisette
Height: Up to 5 metres

It is often said that you can measure the hardiness of your climate by whether this rose will grow well for you. We are fortunate that it does very well for us, although we keep it as a large shrub rather than growing it as a climber. Probably the best-known plant of 'La Marque' that I know is in the garden of the rose authority, the late Nancy Steen in Remuera, Auckland. It grows on a boundary fence and is suspended into a nearby tree at a suitable height to walk under so that the blossoms can be seen to advantage. The large, loose blooms are pure white with pale lemon-yellow in the middle. They are very fragrant and quite double, and usually appear in groups of three to five.

'Lavender Lassie'

Raised by Kordes, Germany
Parentage unknown
Introduced 1959
Type: Shrub-climber
Height: Up to 4 metres

Occasionally this rose is described as a Hybrid Musk but it seems to fit more easily with the classification given above. It is a vigorous grower — sometimes the long shoots reach at least 4 metres. It can easily be grown as a climber or a shrub. The colour is difficult to describe and under my local conditions does not see much of the lavender mentioned in its name. It may be that in another place at another time it can be quite lavender. At best it is pink with the lightest of mauve. The flowers are of medium size and fragrant. They are quite double and plentiful and the plant flowers repeatedly. This is an excellent choice if you are looking for something a little different to vary your planting scheme.

'La Ville de Bruxelles'

Raised by Vibert, France
Parentage unknown
Introduced 1849
Type: Damask
Height: 1.5 metres

When thinking of the beautiful Damask roses, this one must be high up on the favourite list. Everything about it is large but not excessive. The plant grows vigorously with large, light green leaves. The blooms are large, very double and very fragrant. They are an even, rich pink all over with no fading, at least in the early stages of development, and later they reflex towards the outer edge and show a button eye. This is a sumptuous flower in every way.

'Leander'

Raised by David Austin, United Kingdom
Parentage: 'Charles Austin' seedling
Introduced 1982
Type: English Rose
Height: 3 metres

It would be true to say that when a rose is introduced it sometimes does not live up to expectation. On the other hand, sometimes a rose comes before the public with lesser publicity, and with the passage of time turns out to be a very good variety that becomes extremely popular. 'Leander' is such a rose, and it appears to improve with time. It is a tall grower and can be trained as a climber or a shrub. The blooms are medium sized (about 7–8 centimetres wide) and appear in clusters. The colour is basically apricot but with shades of pink, cream and salmon. The flowers open out to a neat rosette.

'Leda' syn. 'Painted Damask'
Raiser unknown; of English origin
Parentage unknown
Introduced prior to 1827
Type: Damask
Height: 1.5 metres

This is a rounded, compact plant with strong, healthy foliage. The buds are reddish-brown and almost always appear to have been chewed by something. However, they eventually end up as a very double, pure white flower with a neat button eye. The reddish colour from the buds appears on the edge of the petals of the blooms in what could be a picotee effect, which gives credence to its common name of 'Painted Damask', as if the edge of the white petals have been brushed with red. The blooms are nicely scented.

'Leontine Gervais'
Raised by Barbier, France
Parentage: R. wichuraiana x 'Souvenir de Catherine Guillot'
Introduced 1903
Type: Rambler
Height: 5 metres

'Leontine Gervais' is a vigorous rambler with slender stems and branches and long wiry growths with shining foliage and deep bronzy young shoots. It is an attractive plant even without flowers. The blooms are sweetly scented, medium sized, cupped at first and opening to semi-double, palish salmon-orange flowers, ageing to a creamy yellow. They arrive in clusters and make a fine display in mid-summer, with some intermittent flowers later.

'Mme Alfred Carrière'

Raised by Schwartz, France
Parentage unknown
Introduced 1879
Type: Noisette
Height: 5 metres or more

There seems to be some doubt about the exact classification of this rose, but it is most often placed in the Noisette group. It is possible that if we did discover the parents we might be very surprised. It is my feeling that we do not see this rose at its best in New Zealand. At Mottisfont Abbey in the South of England, there is an extremely fine specimen growing against a brick wall. It is at least 5 metres wide and up to 3 metres tall. During mid-summer it is a sight to behold with its large, fragrant, double white blooms with just a touch of pink at first. The plant is tough, healthy and vigorous, and can be used for several different purposes.

'Mme Alice Garnier'

Raised by Fauque, France
Parentage: R. wichuraiana x 'Mme Charles'
Type: Rambler
Height: 5 metres

'Mme Alice Garnier' is very sweetly scented with small, very double, quilled blooms of medium to soft pink, sometimes with apricot in the middle. The foliage is dark, glossy, abundant and healthy, with bronze young shoots. In some ways, this rose could be likened to a fatter 'Cécile Brunner', but at other times, especially when open, it does not resemble it at all. This is an important acquisition to the rambler group that thrives in most positions.

'Mme Hardy'
Raised by Hardy, France
Parentage unknown
Introduced 1832
Type: Damask
Height: 1.5 metres

This superlative rose has been written about many times. The purest of white flowers develop from fat, creamy-white or blush-pink buds. They are lemon scented, medium sized, cupped, very double, and quartered with a green button eye. It has been said that this is the most exquisite rose of all, and that as one of the classic old roses few others can come near it for sheer perfection. Praise indeed, but then again, it deserves it.

'Mme Jules Thibaud'
Raiser unknown
Parentage: Sport from 'Cécile Brunner'
Introduction date unknown
Type: Polyantha
Height: 1 metre

Many people thought that this rose was lost, but it came to light in Otago, New Zealand, some years ago. It forms the fourth of the group that includes 'Cécile Brunner', 'Perle d'Or', 'White Cécile Brunner' and 'Mme Jules Thibaud'. In fact, there is a fifth member called 'Pasadena Tournament', also known as the 'Red Cécile Brunner'. 'Mme Jules Thibaud' has foliage and growth similar to the others, but the flowers have a few more petals, and there are times when they are quite deep pink, especially when newly opened. Sometimes there is a little orange in the throat, which pales with age. Because of the manner of the new season's growth (like the other members) it is difficult to obtain sufficient budwood to create enough plants to meet the demand.

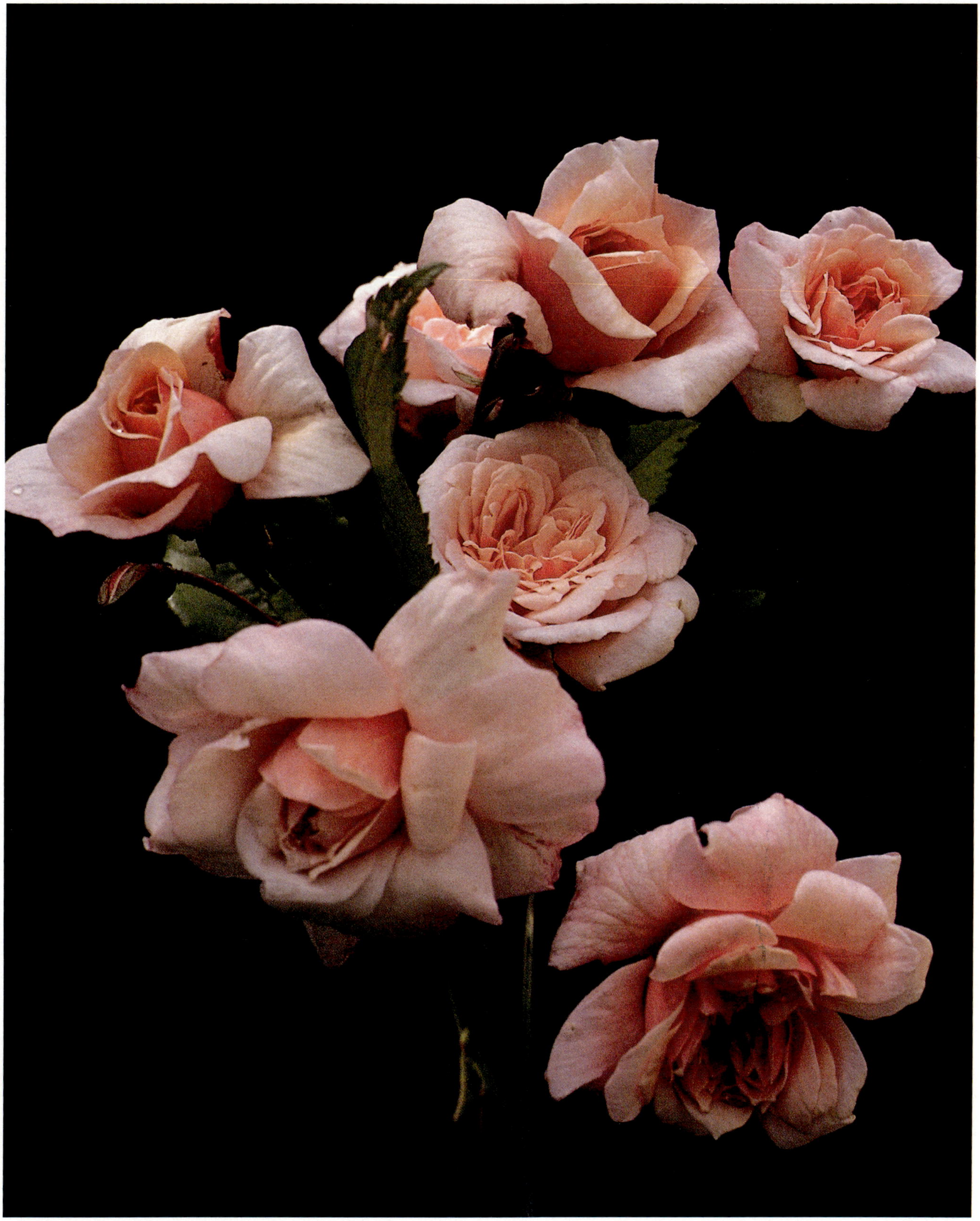

'Magenta'

Raised by Kordes, Germany
Parentage: Yellow Floribunda seedling x 'Lavender Pinocchio'
Introduced 1954
Type: Shrub
Height: 1.5 metres

Although this rose is a reasonably recent introduction, it gains a place in this volume because of its old form. The growth is compact and, as for most of the shrub roses, about as wide as it is tall. The medium-sized blooms of about 8 centimetres are freely produced from early summer onwards, and are at first a lovely shade of deepish lilac-mauve, paling a little with age. They have a good scent and are quite double, appearing on arching branches in lovely sprays that are useful for floral arrangements.

'Magnifica'

Raised by Dingee and Conrad, USA
Parentage: R. rugosa x 'Victor Hugo'
Introduced 1909
Type: Rugosa
Height: 1.5 metres

As a whole, the Rugosas have had rather a raw deal from some rose enthusiasts. It seems to be a matter of acceptance. It was perhaps difficult for some to become used to the idea that these roses actually do perform as beautifully as the other types. In fact, most have one or two additional attributes. First, they can support beautiful, fat, orange fruit while the plant is still in flower and, secondly, almost all of the group grow excellently near the sea. This particular variety is well named, for it is truly magnificent when in flower. The large (12–14 centimetre) blooms are the most beautiful shade of purple, and are double and very fragrant. The plant has handsome green foliage.

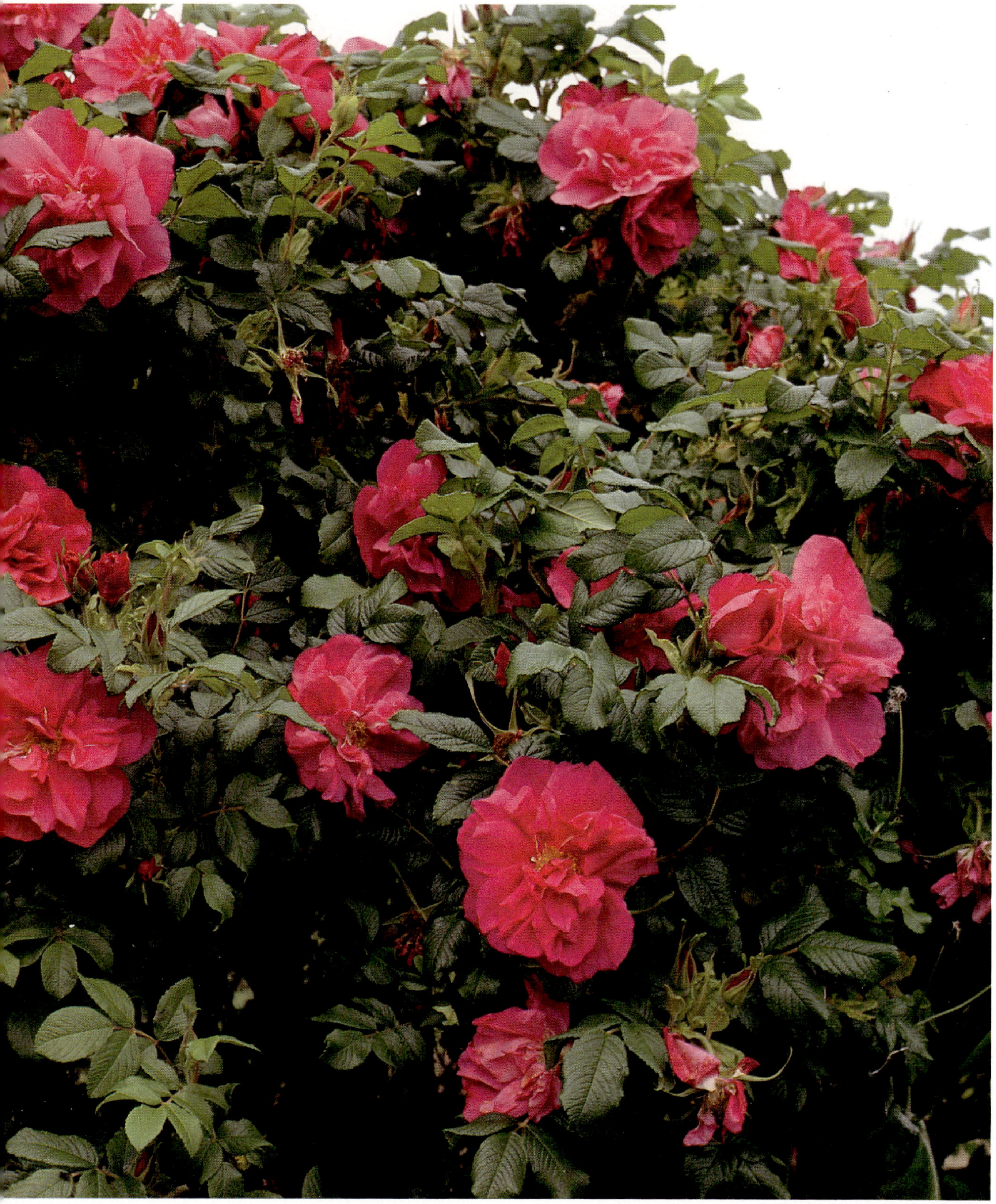

'Maigold'

Raised by Kordes, Germany
Parentage: 'Poulsen's Pink' x 'Frühlingstag'
Introduced 1953
Type: Shrub-climber
Height: 3 metres

Once again we have a beautiful product from Kordes, which has the uncanny knack of introducing very fine roses with great regularity. The truth is probably that having existed through more than four generations, they have created a base from which they are able to be very selective in those they release. A few years ago, Wilhelm Kordes IV told me that they can grow 200,000 seedlings to select only two for one year's introduction. 'Maigold', by nature of its many prickles and distinctive foliage, shows its relationship to *R. spinosissima* through 'Frühlingstag'. The flowers are large, double, very fragrant, and a lovely shade of apricot-orange. It has one massive summer-flowering with a few blooms later in the season.

'Marie Pavie'

Raised by Alegatiere, France
Parentage unknown
Introduced 1888
Type: Polyantha
Height: A little over 1 metre

Polyanthas could probably be divided again and again into sub-branches, because many of those that are placed under the protective umbrella of the Polyantha clan really are quite different from each other. However, having accepted the wider grouping, this member stands out for several reasons. The plant is quite handsome, and if left to its own resources reaches over a metre tall and wide. The foliage is large and bronzy, and seems to be impervious to disease. The buds are fat and arrive in clusters, bursting into the prettiest blush-pink blooms of a rosette nature, about 4 centimetres wide. Deeper at first and paling later, they have a lingering fragrance. This specimen is very useful for tubs and small gardens.

'Mary Rose'
Raised by David Austin, United Kingdom
Parentage: 'Wife of Bath' x 'The Miller'
Introduced 1983
Type: English Rose
Height: 1.5 metres

Looking back ten years, it seems unlikely that two such beautiful roses as 'Mary Rose' and 'Graham Thomas' should have been introduced side by side at the Chelsea Flower Show in 1983. These two excellent roses announced that the English Roses had arrived. 'Mary Rose' brought the Damask Roses into the twentieth century. To all intents and purposes, this lovely hybrid is an ever-blooming Damask that under favourable climatic conditions can literally be in flower continuously. It is a large, rounded shrub with medium to large blooms of bright pink that have a lovely fragrance. This rose was named after Henry VIII's flagship, which was recovered from the Solent after more than 400 years at the bottom of the sea.

'Maxima' — Jacobite Rose
Raiser unknown
Parentage: Possibly related to 'Maiden's Blush'
Introduction date unknown but of great antiquity
Type: Alba
Height: 2 metres

The large flowers of this splendid, classic old rose are creamy-white with muddled centres and are quite fragrant. They are sparingly arranged on a tall plant among the lush grey-green foliage. It has been said that this was the rose used as an emblem by Bonnie Prince Charlie and his followers. One has to ask whether a rose that flowers for about two months of the year would have been useful in this way. It is tough, and will grow well in the coldest climates.

'Mermaid'

Raised by Paul, United Kingdom
Parentage: R. bracteata x 'Double Yellow Tea Rose'
Introduced 1918
Type: Climber
Height: 8 to 10 metres

Here we have what might be called a classic old climber. *R. bracteata*, a distinctive rose in its own right, helped to create the beautiful 'Mermaid'. The flowers are large, all of 13–14 centimetres, with just five single petals. When fully open they expose stamens of a bronze colouring that remain part of the plant's display even after the petals fall. The blooms are golden-yellow and have a strong fragrance. The growth can be very vigorous and the stems are well covered with large, hooked thorns. 'Mermaid' requires some shelter for success as it is not totally hardy. It also requires quite a lot of room and so is not suitable for small areas.

'Moonbeam'

Raised by David Austin, United Kingdom
Parentage unknown
Introduced 1983
Type: English Rose
Height: 2 metres

If you are looking for a rose with great delicacy and extreme beauty, and select the right sheltered spot, you will love this hybrid. It has tall, upright growth, freely producing large, double flowers. They are about 14 centimetres wide and just a little more than double. The creamy, long, pointed buds with the barest flush of apricot open to a lovely flower with a delicate scent. 'Moonbeam' is definitely a collector's item.

'Moonlight'

Raised by Pemberton, United Kingdom
Parentage: 'Trier' x 'Sulphurea'
Introduced 1913
Type: Hybrid Musk
Height: 2 metres

This is another of the Rev Joseph Pemberton's beauties, with Peter Lambert's 'Trier' lurking in the background. 'Moonlight' has small to medium flowers, almost single, opening wide in clusters. They are yellowish in the bud and open to creamy-white with prominent golden-yellow stamens. The blooms are striking among the glossy, dark, bronzy foliage. The other parent of 'Moonlight' was 'Sulphurea', a yellow Hybrid Tea introduced by W. Paul in 1900.

'Moyesii Geranium'

Raised by Mulligan, United Kingdom
Parentage: Original form R. moyesii *from Western China*
Introduced 1938
Type: Species hybrid
Height: 2.5 metres

Whenever a rose is sought after for hips or fruit in the autumn, inevitably the moyesii group come into the discussion. You may have read that 'Geranium' is the shortest among them, but remember that this means at least 2 metres. It is unquestionably a most beautiful and interesting rose. The medium-sized, single blooms are a most intense orange-scarlet. They have attractive stamens that are very prominent. When the blooms pass, the fruit develops into long (8–9 centimetres) flagon-shaped beauties. They are at first lime-green in late summer, and as the weather cools they turn fiery orange and make a magnificent sight.

'New Dawn'
Raised by Somerset Rose Nursery, USA
Parentage: 'Dr W. Van Fleet' *sport*
Introduced 1930
Type: Climber
Height: 5 metres or more

In every group or sub-group of the rose family there are several varieties that stand out from the rest. They do so because they are 'happy roses'. These are the ones that thrive in most places without any problems. 'New Dawn' is a 'happy rose', and not only has it become very popular around the world, but it has passed its genes on to many other superlative varieties. The sweetly scented flowers are medium to large with a little more than a blush of pink when open. They are lightly double and open wide, showing golden stamens. The growth is almost rambling, with long, wispy branches that can be easily trained. Altogether it is an excellent rose.

'Nuits de Young' syn. 'Old Black'
Raised by Laffay, France
Parentage unknown
Introduced 1845
Type: Moss
Height: 1 metre

'Nuits de Young' is considered by many to be the darkest and most beautiful of the Moss roses. It grows into a tidy, compact plant that during the summer season sports a lovely display of small to medium double blooms of the deepest crimson-maroon and rich purple with a few golden stamens. I have seen another variety that may be a little different in that the petals open wider, showing a fine group of golden stamens. In Denmark it is known as 'Old Black'.

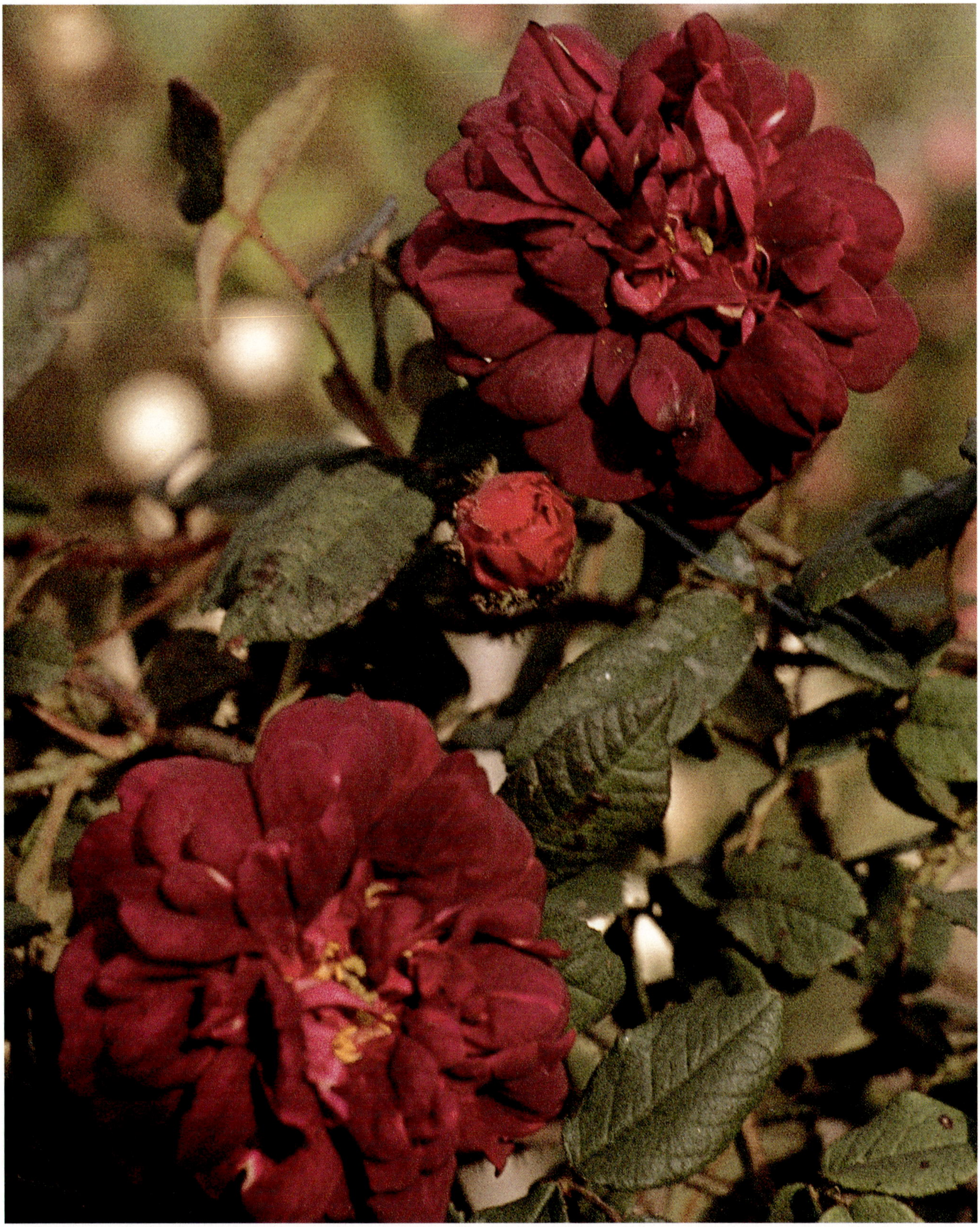

'Old Blush China'

Raiser and parentage unknown
Introduction date unknown but of great antiquity
Type: China Rose
Height: 1.5 metres

It is interesting that two very old roses should follow each other. Although they come from two different continents and each in its own way is so diverse, they have much in common. As with the former, this rose has a number of alternative names, indicating that at different periods of time, in different countries, this rose was known and loved. It was introduced into Europe about 1750, but there is no doubt that it was known in China for a very long time. It was part of the gene pool that came to the Western world from the Far East and in a peaceful revolution changed the summer-flowering-only roses into all-season bloomings. The rose-pink, loosely double flowers of 'Old Blush China' are medium sized and appear from summer through to winter. There is a climbing form of this fine old variety that can easily reach up to 15 metres.

'Parkdirektor Riggers'

Raised by Kordes, Germany
Parentage: R. kordesii x 'Our Princess'
Introduced 1957
Type: Shrub-climber
Height: 4 metres or more

This excellent rose has the inbuilt ability to perform well. Its flowers are deep, velvety crimson, a little more than single, and of medium size, about 8 centimetres. They open out flat and the petals are wavy. As many as forty to fifty blooms can appear on one large head. The flowering period extends from early summer to late autumn. This hybrid is always favourably commented upon in our display garden.

'Paul Transon'
Raised by Barbier, France
Parentage: R. wichuraiana x 'L'Ideal'
Introduced 1900
Type: Rambler
Height: 5 metres or more

'Paul Transon' is a happy rose that does well in most situations. It can be a rampant grower with long canes. The foliage is dark and glossy and the young growth bronzy and dark, which makes a lovely setting for the medium-sized (8–10 centimetre), sweetly fragrant blooms. The flowers are a little unusual in that the petals appear quilled. The colour is a salmon-pink, deeper in the centre at first, paling towards the edges later. This is a beautiful, useful and attractive variety that could find a place in every garden.

'Penelope'
Raised by Pemberton, United Kingdom
Parentage: 'Ophelia' x 'William Allen Richardson' or 'Trier'
Introduced 1924
Type: Hybrid Musk
Height: 2 metres

Despite the parents given above for this fine rose, I can see no 'William Allen Richardson' in it at all. 'Trier' is nearer the mark as on close examination it is not unlike 'Penelope'. With hindsight, 'Penelope' helped to found the Pemberton dynasty of Hybrid Musks. The 'man of the cloth' became the 'man of the rose'. 'Penelope' has medium-sized, semi-double blooms that open pale pink but soon pass to creamy-white when fully open and flat. They are fragrant, and the display of blooms looks very attractive on a healthy plant and lasts from summer to autumn.

'Perdita'

Raised by David Austin, United Kingdom
Parentage: 'The Friar' x (Unnamed seedling x 'Iceberg')
Introduced 1983
Type: English Rose
Height: 1.5 metres

This was one of what David Austin regards as his first flush of English Roses to appear on the scene before the introduction of 'Graham Thomas' and 'Mary Rose' in 1986, which marked a new development in the English Rose. The blooms are large and many petalled and have a strong fragrance. They are apricot in the bud and open paler to a cupped flower with an old look about them. The scent must be very good, because 'Perdita' was awarded the Royal National Rose Society's Henry Edland Medal for fragrance in 1984. This is a healthy plant with a spreading habit.

'Perle d'Or'

Raised by Rambaud, France
Parentage: Polyantha x 'Mme Falcot'
Introduced 1884
Type: Polyantha
Height: 1 metre

'Perle d'Or' is another member of the 'Cécile Brunner' group, which has delighted gardeners for many years. One has to wonder why such excellent hybridists have taken so long to turn their attention towards creating more of this type. 'Perle d'Or' is apricot-yellow in the bud, opening to yellow and paling towards the edges. In every other respect it is typical of the group and has a lovely scent.

'Phyllis Bide'
Raised by Bide, United Kingdom
Parentage: 'Perle d'Or' x 'Gloire de Dijon'
Introduced 1923
Type: Rambler
Height: Up to 5 metres

It seems prophetic that the very next rose should have 'Perle d'Or' as one of its parents. 'Phyllis Bide' is one of the nicest happenings in the rose kingdom. It is a 'happy rose' and has the ability to flower for a very long time, even into the middle of winter in sheltered places. An individual bloom would never set the world on fire, but collectively a large spray is extremely beautiful. The flower is small, about 3 centimetres, but when grouped, especially with the changes of colour from lemon-yellow in the bud to pink, from pink to rosy-pink, paling to shades of apricot and cream, it becomes a fine sight indeed.

'Pinkie'
Raised by Swim, USA
Parentage: 'China Doll' *open pollinated*
Introduced 1947
Type: Polyantha and Climber
Height: 1 metre and 5 metres

Like its parent, this rose has a great abundance of bloom, making it very showy. The colour is the brightest pink with a faint trace of cerise mixed in. The flowers are 7–8 centimetres wide, loosely semi-double and have a light scent. The climbing type is perhaps more attractive because of the great density of flowers on a much larger plant. Provided you like the colour, it is well worth a place in any garden.

'Poulsen's Crimson'

Raiser and origin unknown
Parentage: 'Orange Triumph' x ('Betty Uprichard' x 'Johanniszauber')
Introduced 1950
Type: Shrub
Height: 1 metre or more

If you admire this rose you may, as I do, ask why hybridists continue chasing the elusive dream when they already have in their hands the most wonderful creations. Often their efforts result in roses that are very near existing ones. You may also ask how such a beautiful crimson rose comes from the above parents, but when we realise that the 'Johanniszauber' part of the equation has 'Château de Clos Vougeot' blood in its veins then all is clear. The beautiful crimson, almost single blooms are 5 centimetres wide with a light scent and appear in large clusters.

'Portland Rose'

Raiser and origin unknown
Introduced about 1800
Type: Portland
Height: 1 metre

Shrouded in the mists of time, we are left to deduce the origins of this rose and the rest of the group. The China Roses brought the recurrent gene to Europe and this was the first group to be developed with their influence. They formed a bridge between the classical old roses with the once-flowering, summer-only habit, and what was yet to come. However, while the Portland Roses were soon passed over, there were some very beautiful members of the group. This rose could be likened to the 'Red Rose of Lancaster' except that it has seasonal flowerings. It has large, semi-double, scarlet blooms with yellow stamens and quite a strong Damask scent.

'Pretty Jessica'
Raised by David Austin, United Kingdom
Parentage: 'Wife of Bath' x unnamed seedling
Introduced 1983
Type: English Rose
Height: 1 metre

'Pretty Jessica' is a beautiful, small-growing rose that could be placed in any small garden. It has the look of a Centifolia, the strong old-rose fragrance of yesteryear, and bright, rich pink double blooms. For most of the season, what more could one want? The growth of the plant is upright, and the foliage reasonably healthy. The name, too, has a bearing on the popularity of this rose. It is amazing how many people have Jessicas in their families.

'Robin Hood'
Raised by Pemberton, United Kingdom
Parentage: Seedling x 'Miss Edith Cavell'
Introduced 1927
Type: Hybrid Musk
Height: 1.5 metres

Generally accepted as a Hybrid Musk, this rose is so bright and so packed with flowers that it makes everything near it look sombre. This is not to say that 'Robin Hood' is not a beautiful rose in its own right, but its brightness is perhaps not consistent with the more reserved shades of the genuine old roses. Its small, bright cherry-red blooms are so tightly packed on the plant at full flower that you cannot see any foliage at all. It is a tough shrub, suitable for many purposes.

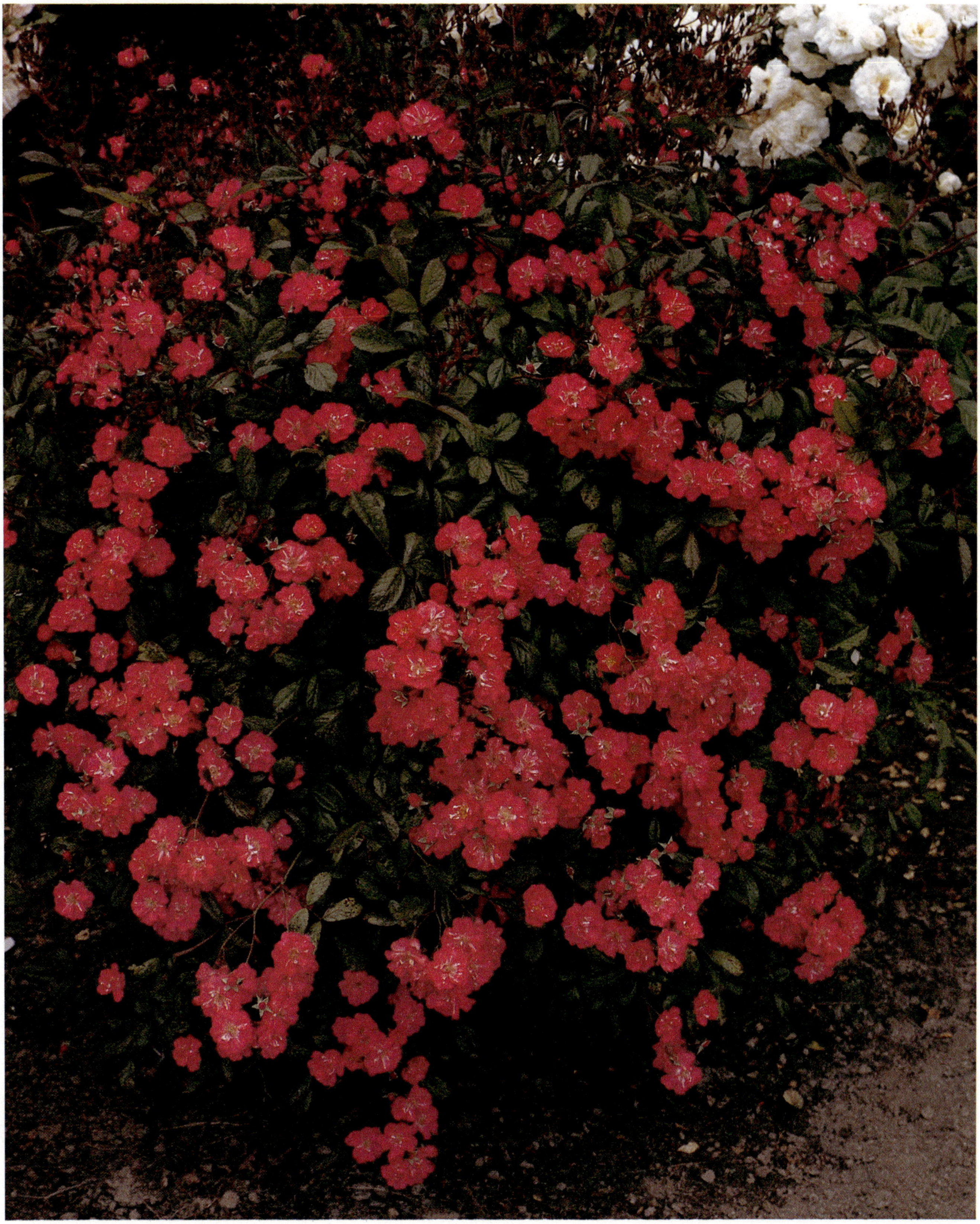

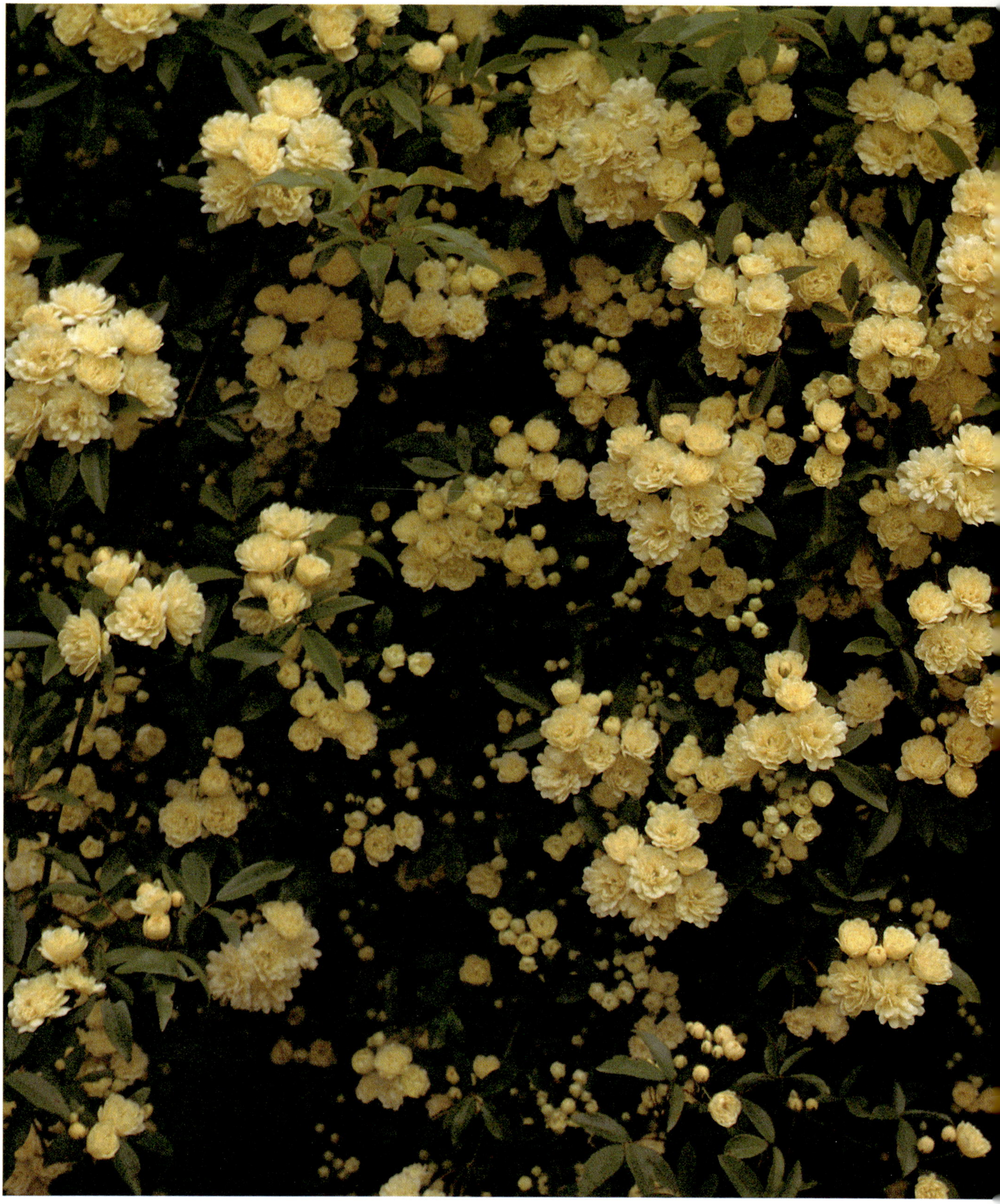

Rosa banksiae lutea
Origin unknown; native of China
Introduced 1824
Type: Species
Height: At least 5 metres

The introduction date given here is misleading because this rose was probably known in China for thousands of years before it was taken to Europe. However, it is a much-respected and desired member of the species group. The plant is very vigorous in milder climates but can suffer frost damage in colder areas. Some retail outlets advertise it as hardy, which is not true and does mislead people. The flowers of this beautiful rose are tiny but prolific, appearing in clusters. The season extends for almost two months, from the moment the first tiny clusters break out on the sunny side of the plant, until the last ones complete the display on the shady side. The branches are thornless and although it is said to be evergreen, this is true only in the warmer climates. Any scent is difficult to detect.

Rosa banksiae lutescens

Raiser unknown
Parentage unknown; native of China
Introduced 1870
Type: Species
Height: At least 5 metres

There are four Banksia roses in cultivation: the double yellow *R. banksiae lutea*, the single yellow *R. banksiae lutescens*, a double white and a single white. There are also two hybrids raised in Italy, both of which are now widely available. This rose, *R. banksiae lutescens*, has a lovely fragrance and is deep golden-yellow in colour. It is a true single, having only five petals with prominent yellow stamens. The blooms appear in large clusters and are less than 3 centimetres wide. The four introduced species of *R. banksiae* are similar in growth and foliage, with only minor differences in twigginess and colour of foliage.

'Rosa Mundi' syn. *Rosa gallica versicolor*

Parentage: Colour sport from R. gallica officinalis
Introduced prior to the sixteenth century
Type: Gallica
Height: 1 metre

The term 'sport' or mutation comes from a twist of nature. All plants have the ability to change a portion of their flower or foliage to another form or colour. This is how many new varieties are created, especially within the genus *Rosa*. Some people do not understand this and may blame the grower when a white rose such as 'Sneprincesse' suddenly sends out a branch of bright orange blooms, which of course was its parent. 'Rosa Mundi' is a sport from the 'Apothecary's Rose' (*R. gallica officinalis*) and is identical except for its colour, which is basically pink with splashes of white, crimson and purplish-pink. The attraction of this rose may be its irregular markings, wherein lies its beauty.

Rosa x richardii

Parentage: R. gallica x R. phoenicia
Introduced to Europe 1902; ancient origin
Type: Species
Height: 1 metre or more

Here we have a rose steeped in history. Its suggested parents are ancient and quite important historically. *R. richardii* is known under several names. It was formerly known as *R. sancta* but it is also known as 'Abyssinian Rose', 'Rose of the Tombs', 'St John's Rose', and 'Holy Rose'. During the building of the Aswan Dam, the Temple of Abu Simbel had to be relocated or lost under the new lake. During the shift, several ancient mummies were opened and, in one, the petals and pollen grains of a rose were identified as *R. richardii* and carbon dated at 2600 BC. True or false, this rose is ancient. Its flowers are large (about 10 centimetres) and single. The petals are crinkled and pale to medium pink. The growth of the plant is compact and it is quite hardy.

'Roseraie de l'Hay'

Raised by Cochet-Cochet, France
Parentage: Possibly a sport from R. rugosa
Introduced 1901
Type: Rugosa
Height: Up to 3 metres

'Roseraie de l'Hay' is a magnificent shrub with deep green foliage and evenly spaced blooms covering the mounded plant. The flowers are a purplish-crimson and large, about 12–13 centimetres wide. They are quite double and fragrant, but unlike most of the Rugosa group, do not really set fruit.

'Rote Max Graf'

Raised by Kordes, Germany
Introduced 1980
Type: Procumbent Climber
Height: 3 metres or more

This comparatively new procumbent rose has the most brilliant, crimson-scarlet, single flowers you could wish for. They appear in clusters and are about 6–7 centimetres wide, paling a little towards the centre. The plant develops long, arching branches, and when in full flower is a sight to behold. It has one long flowering season lasting about two months and falling between those that flower early and again later. This rose can be trained to climb or ramble and could just as easily fall over a bank or terrace.

'Rugspin'

Raised by Petersen, Denmark
Parentage: Rugosa seedling
Introduction date unknown
Height: 2 metres

For many years I was able to import beautiful roses from my old friend Valdemar Petersen of Love in Denmark. His roses were always accurately named and always arrived in excellent condition. He sometimes slipped one of his own creations in the parcel, and this was how 'Rugspin' arrived in New Zealand. It is truly a beautiful rose. The blooms are 13–14 centimetres wide and the five single petals are royal purple in colour, followed by large, fat, orange fruit. It is a fragrant rose that will remain a lasting memory to a fine, dedicated rosarian.

'Sally Holmes'
Raised by Holmes, United Kingdom
Parentage: 'Ivory Fashion' x 'Ballerina'
Introduced 1976
Type: Hybrid Musk
Height: 2 metres

Some people may raise their eyebrows at this rose being placed among the Hybrid Musks, but it has just as much right to be placed with them as anywhere else. The large, single flowers appear in clusters and the plants are literally choked with blooms. Creamy-white at first, they become whiter and are fragrant. It has been said that there are too many blooms on the heads and this tends to make the branches blow around somewhat. No doubt this rose will be used for future breeding as it seems to be strong in all facets of the hybridisers' requirements.

'Sanders' White'
Raised by Sanders, United Kingdom
Parentage unknown
Introduced 1912
Type: Rambler
Height: 5 metres

Over a long period of time a white rambler with tight little rosette flowers has been grown in New Zealand. It was known as 'Sanders' White'. But after three journeys to the United Kingdom and Europe, I realised that the rose grown in these countries under that name was different. We now grow that rose, and it has slightly bigger, pure white blooms that are not quite so double as the former one, and it is possible to see golden stamens in the centre. The new version is fragrant and has deep green glossy foliage that contrasts beautifully with the white flowers. It blooms from summer through to autumn, performs well as a rambler, and is excellent trained on a pole or as a weeping standard.

'Sea Foam'

Raised by Schwartz, USA
Parentage: 'White Dawn' x 'Pinocchio'
Introduced 1964
Type: Shrub-climber
Height: 3 metres

The parentage given above is not accurate but it does involve the same parents at least three times. This rose seems to be in no man's land between a shrub and a climber. It makes an excellent shrub that spreads and gains height with age, while if trained to climb it will do so, although it certainly takes time to achieve its object. However, it is an excellent rose — tough, wiry, procumbent if needs be, prolific in flower and lightly scented. The blooms are 3–4 centimetres wide, developing from fat, pink buds, usually opening blush-pink and with a little age becoming creamy-white. They are quite double, rosette type, and are packed tightly on the plant, blooming from summer to autumn.

'Semi-plena'

Raiser unknown
Parentage: Sport of R. alba maxima
Type: Alba
Height: At least 2 metres

Graham Thomas, the doyen of old roses, writes: 'The White Roses (the Albas) are supreme over all the other old races in vigour, longevity, foliage, delicacy of colour, for they embrace some exquisite pink varieties, and purity of scent.' 'Semi-plena' is a fine tribute to an excellent group. Its toughness, durability and disease resistance come from its relationship with *Rosa canina*. This rose is believed to be the 'White Rose of York'. It has grey-green, healthy foliage, medium to large, almost single blooms with a good fragrance. It is one of several in the group from which Attar of Roses is distilled.

'Semperflorens' syn. 'Slater's Crimson China'

Origin unknown
Parentage: R. chinensis *derivative*
Introduced 1792
Type: China Rose
Height: About 1 metre

This very bright crimson, small, cupped rose grows quietly in many gardens, and I am sure the owners do not realise how important it has been. No one knows for how many centuries this and its near relatives, 'Fabvier' and 'Cramoisi Superieure' grew in Chinese gardens, but when it was introduced into England in the late eighteenth century, it brought about a peaceful revolution in roses on two fronts. First, it brought its lovely colour, which really had not been seen in that part of the world before. Secondly, it introduced the recurrent gene into roses, which created many new groups within the genus. This historic rose deserves more recognition.

'Sissinghurst Castle'
Origin and date of introduction unknown
Type: Gallica
Height: At least 1 metre

When one thinks of the stories about the roses raised by French growers in the early nineteenth century, and how many Gallicas were among them, it is amazing that many of them have survived to this day; and it is even more amazing that accurate names have accompanied them. One of these roses is 'Sissinghurst Castle', the name given to it by Harold Nicholson and Vita Sackville-West when they discovered it in the grounds of their home in the 1940s. More recently, it is thought to be 'Rose des Maures'. The plum-coloured, medium-sized flowers are fragrant and have muddled centres showing some stamens.

'Sombreuil'
Raised by Robert, France
Parentage: 'Gigantesque' seedling
Introduced 1850
Type: Climbing Tea Rose
Height: At least 4 metres

Often when customers ask for an old climbing rose it takes a little time to establish exactly what they are looking for, and after this has been done, more often than not this is the rose chosen — especially if it happens to be in flower at the time. It has a very good tea scent and the open flower is large (about 10 to 12 centimetres) and literally packed with petals that appear to be quilled and come from the centre. It opens very flat and is creamy-white. There always seems to be a flower on the plant, which has long branches, good foliage and large thorns. It is an excellent variety for every reason.

'Souvenir de la Malmaison'
Raised by Beluze, France
Parentage: 'Mme Desprez' x a Tea Rose
Introduced 1843
Type: Bourbon
Height: 1 metre

'Souvenir de la Malmaison' is probably the first old rose that comes into people's minds, even when they do not know its name. This wonderful variety conjures up the best features for those who are new to the older-type roses. It seems to have everything. The fragrant flowers are large — at least 12 centimetres across — well-filled with petals and nicely quartered. The colour is an attractive shade of pink across most of the bloom except near the edges, where it becomes creamy. The flower develops from a fat, rounded bud, and when open is very flat across the top. A good climbing form is available and is very popular.

'Sparrieshoop'
Raised by Kordes, Germany
Parentage: ('Baby Chateau' x 'Else Poulsen') x 'Magnifica'
Introduced 1953
Type: Shrub-climber
Height: At least 4 metres

The House of Kordes in Holstein, Germany, must have thought a great deal of this rose to have named it after the district in which their nursery is situated. This is a hardy plant with strong, tall branches and large thorns. The blooms are 10 centimetres across and have shades of pink, apricot and salmon, and are paler in the centre. They are a little more than single and nicely fragrant, and arrive in clusters. The healthy plant seems to be in flower almost continuously.

'Stanwell Perpetual'

Raised by Lee, United Kingdom
Parentage: R. damascena semperflorens x R. spinosissima
Introduced 1838
Type: Species hybrid
Height: Over 2 metres

This rose originated as a chance seedling in a garden at Stanwell in Essex. In many well-established gardens rose seedlings have a habit of popping up anywhere, and most of them amount to nothing of importance, but sometimes one turns out to be a beauty. That is the case with this chance hybrid. It is a superlative rose in every way. When Mr Lee named it 'Perpetual' he recorded probably its finest feature. The large, healthy plant with greyish fern-like foliage really is never without flower. They are of medium size (about 9 centimetres wide), flat and double and a little deeper than blush-pink when opening, paling later.

'Sweet Juliet'

Raised by David Austin, United Kingdom
Parentage: 'Graham Thomas' x 'Admired Miranda'
Introduced 1989
Type: English Rose
Height: Up to 2 metres

There is no doubt that when used as a parent 'Graham Thomas' has a strong influence on its progeny. The plant growth and the foliage have been influenced in this way. The upright, healthy growth supports ample blossoms of medium size. They are cup shaped and of a pleasant buff, apricot and yellow colouring — shading that has become very popular in recent years — and they have a strong tea scent.

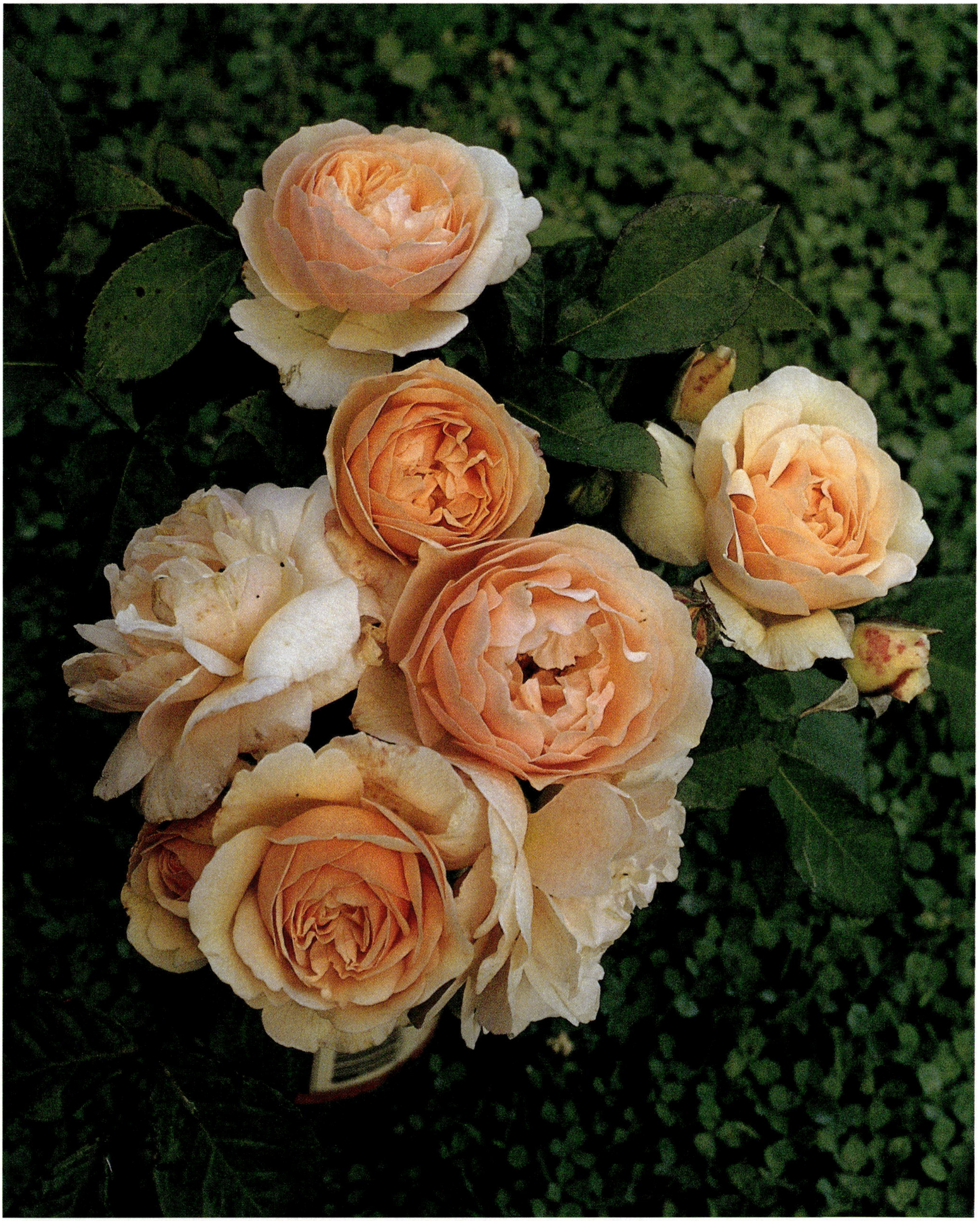

'The Fairy'
Raised by Bentall, United Kingdom
Parentage: 'Lady Godiva' *sport*
Introduced 1932
Type: Polyantha
Height: 1 metre

It is difficult to write about a rose that is so universally popular. No matter which country you visit this rose has been accepted and used by all. It is found almost everywhere. The fact that it is so well used seems to ignore its inherent fault that it has no scent. However, it has many other good features. Probably its popularity comes first from its colour, and secondly from its hardiness and prolific flowering. The colour is bright, clear pink, and the blooms, although small, arrive in clusters, literally covering the plant in full bloom. It is extremely hardy and can be used for many purposes including tubs, standards and groundcover.

'The Garland'
Raised by Wells, United Kingdom
Parentage: R. moschata x R. multiflora
Introduced 1835
Type: Rambler
Height: 3 to 4 metres

'The Garland' is a rampant, vigorous grower that has a superlative display in mid-summer only. The 3 centimetre blooms are set out in clusters, and the individual flowers are not unlike a double daisy, billowing out from the centre exposing the stamens. The flowers appear in shades of lemon and pink, paling to white, and of course have a lovely fragrance.

'The Pilgrim'

Raised by David Austin, United Kingdom
Parentage: 'Graham Thomas' x 'Yellow Button'
Introduced 1991
Type: English Rose
Height: 1.5 metres

This new rose could very easily be taken for an old one. The only feature that gives it away is its lemon colour, a shade not present in the classic old roses. The flowers of this beautiful hybrid are large and, when open, very flat and packed with petals not unlike the Tea Rose 'Sombreuil'. It has a strong fragrance and is bound to become popular when well known. No doubt in the future its progeny will have a great influence in this changing world of rose form.

'Tricolore de Flandre'

Raised by Van Houtte, Belgium
Parentage unknown
Introduced 1846
Type: Gallica
Height: Over 1 metre

Another of the very beautiful striped old roses, 'Tricolore de Flandre' is in some ways not unlike 'Camaieux', and is sometimes mistaken for it. This hybrid has a spreading growth habit, and the flowers are of medium size, double and nicely fragrant. The blooms would originally have been white with heavy striping of crimson-purple, purple and lilac-mauve. Some rose enthusiasts have been known to look down on striped roses but they are beautiful and distinctive in their own right and make a marvellous contrast when planted among other colours.

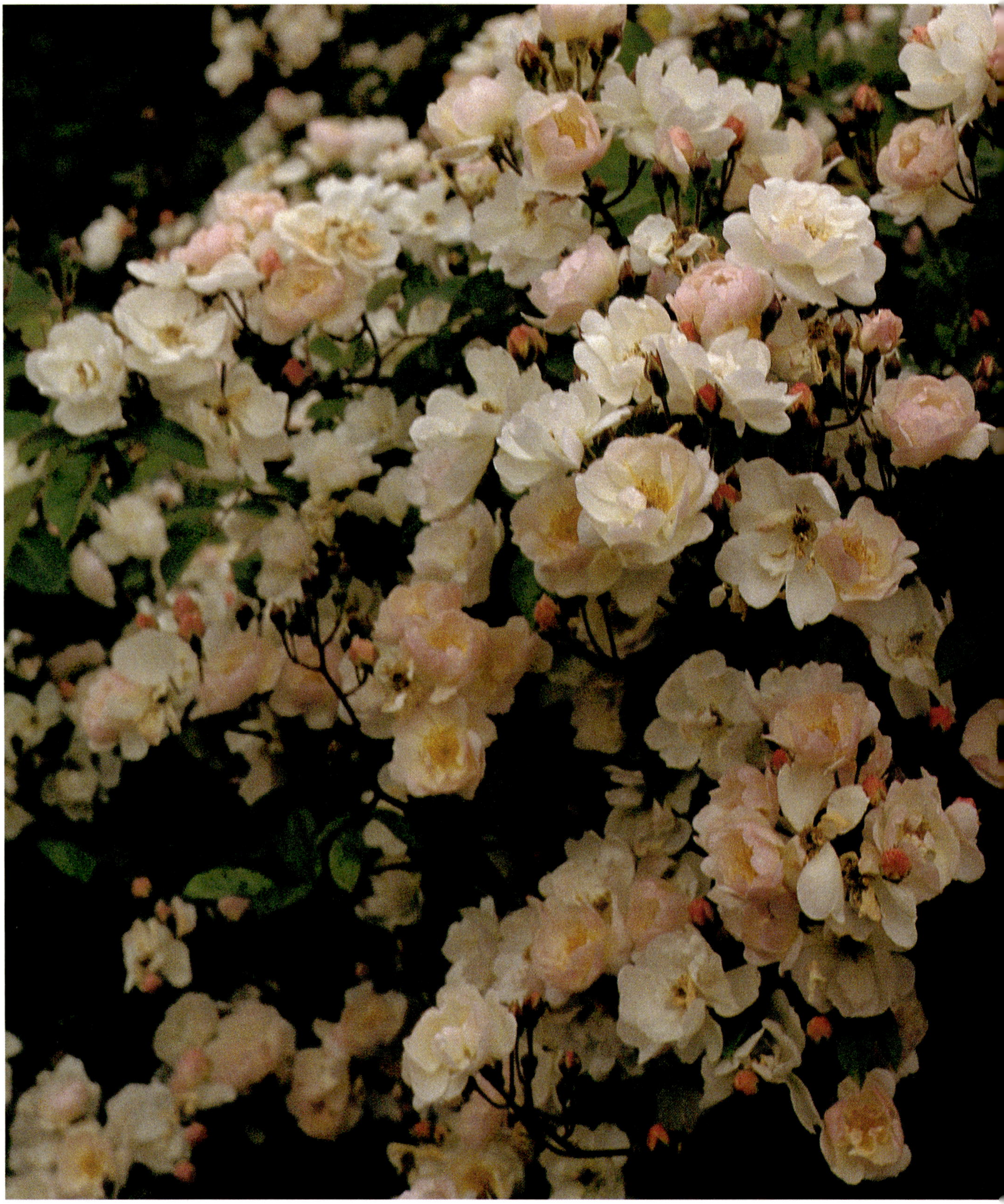

'Trier'

Raised by Lambert, Germany
Parentage: Suggested 'Aglaia' x 'Mrs R. G. Sharman-Crawford'
Introduced 1904
Type: Hybrid Musk
Height: 4 metres or more

Whether the above parentage is correct or not, one has to ask if Peter Lambert ever dreamed that this rose would become so important. It has semi-double, small, fragrant blooms in large clusters. They are lemon-blush at first, becoming blush-pink and white. The two major contributors to the creation and development of the Hybrid Musks were Peter Lambert and Joseph Pemberton, and both used 'Trier' in their respective work. Their resultant efforts speak for themselves, and again we have the example of a humble rose of no great expectation becoming the basis for the dedicated work of two men and the creation of a new rose type.

'Trigintipetala' syn. 'Kazanlik'

Raiser and parentage unknown
Date of introduction unknown, but very old
Type: Damask
Height: 2 metres

Probably more correctly named *R. damascena trigintipetala*, this old variety became the basis of the distillation of Attar of Roses used in the cosmetic and perfume industry. It is a very old industry and one that is still functioning in many countries such as France, Turkey, Egypt, India, Morocco, Spain, Bulgaria and Iran. Several years ago, some 7,500 acres of roses were grown in Bulgaria and the Kazanlik Valley for this purpose, and in recent years the production of Attar is known to have increased. In the old Soviet Union this rose was hybridised with several near relations in an attempt to improve the production of Attar. This rose is a lusty, healthy grower and has medium-sized, medium-pink loose blooms with a typically strong Damask fragrance.

'Tuscany Superb'

Raiser unknown
Parentage: May be a seedling from 'Tuscany'
Introduced possibly before 1848
Type: Gallica
Height: At least 1.5 metres

Few classic old roses command such immediate attention and complimentary comment as this hybrid. A plant in full display covered in medium-sized crimson-purple double blooms is a joy to behold. When fully open a few stamens are visible and there is quite a strong fragrance. The petals are often used in pot pourri.

'Vanity'

Raised by Pemberton, United Kingdom
Parentage: 'Château de Clos Vougeot' x seedling
Introduced 1920
Type: Hybrid Musk
Height: Up to 3 metres

Deep rose-pink, almost single blooms prolifically covering a large plant make 'Vanity' a grand sight. The blooms are at least 10 centimetres across and open out flat and are deliciously fragrant. Perhaps this rose has not achieved the popularity it deserves because of its colour, which some would describe as a rather harsh pink. Nevertheless, it is a fine rose, and can be made to climb quite easily — another excellent product from the Pemberton stable.

'Veilchenblau'
Raised by Schmidt, Germany
Parentage: 'Crimson Rambler' x 'Erinnerung an Brod'
Introduced 1909
Type: Rambler
Height: 5 metres or more

'Veilchenblau' is a much-talked-about and loved rose that is known all over the world. It has practically no thorns at all and for this reason alone is sought after by some rosarians. But it has much more to offer. It is a gentle but consistent grower and in full bloom is densely covered with small (about 3–4 centimetre) semi-double blooms of several different shades: mauve, lilac, grey, purple and a little white. The unusual flowers also have a lovely fragrance. This hybrid, along with 'Bleu Magenta', 'Amethyste', 'Rose Marie Viaud' and 'Violette' form an important group among the ramblers, all of different shades of purple and mauve.

'Viridiflora' — *R. chinensis viridiflora*
Raiser and origin unknown
Introduced before 1855
Type: China Rose
Height: 1 metre

Again we have a rose that comes from China — a country that holds so many mysteries for westerners. We do not know how long many of the roses in this group were known and appreciated in that country, but we must be forever grateful that they have mostly found their way into our gardens. It could be said that this rose is loved by a few and despised by many. It has greenish, rusty-looking arrangements that serve for flowers, and the least you can say is that it is unusual. Roy Rumsey of Dural in New South Wales told me some years ago that several of these flowers suddenly appeared on a red China Rose in his garden. Is this the answer to the mystery?

'Wedding Day'

Raised by Stern, United Kingdom
Parentage: R. sinowilsonii x *unknown rose*
Introduced 1950
Type: Rambler
Height: Up to 10 metres

This is an appropriate name for a rose with white flowers, but it was so named because the flowers first opened on the raiser's wedding anniversary. For twenty years a similar rose has been incorrectly given this name in New Zealand, but the hybrid known in Western Europe as 'Wedding Day' is very different. The genuine one resembles the other but its flowers have a gap between the petals, while the growths and stems are thinner and more wiry. The foliage is a darker green and quite glossy. Until a name can be found for the incorrect one, the other has been called 'English Wedding Day'. Both have small, white, single, fragrant blooms in large clusters followed by tiny orange fruit after a summer-long display.

'White Cécile Brunner'

Raised by Fauque, France
Parentage: 'Cécile Brunner' *sport*
Introduced 1909
Type: Polyantha
Height: 1 metre

As this is a colour sport, the shape and composition of the buds and flowers are very similar to the parent, but the foliage is light green and the colour is, of course, quite different. If we became very critical we could say that this rose is not white but very close to it. The buds are quite yellow in the tight stage, pale to lemon-yellow at first opening and then creamy-white later. The flowers are just as freely produced as in the parent and they are also pleasantly fragrant. This rose is much sought after for floral work.

'White New Dawn' syn. 'Weisse New Dawn'

Raised by Longley, USA
Parentage: 'New Dawn' x 'Lily Pons'
Introduced 1949
Type: Climber
Height: 5 metres or more

'White New Dawn' is a very valuable addition to the climbers. I saw this rose for the first time in 1988 at the State Rosarium at Sangerhausen in East Germany. Needless to say, I was so impressed with it that it was imported into New Zealand and then released for distribution two years later. It has dark green, healthy foliage with large thorns on the long stems and the young growth is quite dark and bronzy. The flowers are fragrant and medium to large with thirty to thirty-five petals. They are reddish in the bud, opening to a milky-white gardenia-like bloom.

'Wife of Bath'

Raised by David Austin, United Kingdom
Parentage: 'Mme Caroline Testout' x ('Ma Perkins' x 'Constance Spry')
Introduced 1969
Type: English Rose
Height: 1.5 metres

Tight little buds develop into cupped blooms of clear pink, paling towards the edges. The blooms are medium sized and when open loosely quartered and have a strong fragrance of myrrh. When Seizo Susuki, the noted Japanese rosarian, visited our garden he found the name of this rose amusing and suggested it might be 'Wife in Bath'. Having had some experience of the difficulties of the English language in other countries, I explained that in his country the rose name could have been 'Wife of Tokyo', and that 'Bath' was an English town, not a bathing vessel as he had at first thought. This rose is a tough, compact, repeat-flowering hybrid suitable for small gardens.

111

'William Lobb'

Raised by Laffay, France
Parentage unknown
Introduced 1855
Type: Moss
Height: Over 2 metres

'William Lobb' has medium to large flowers with muddled centres, purplish-crimson (more purple than crimson) at first, turning to lavender-purple then greyish-lavender with a lighter reverse. The plentiful blooms have a strong scent and look well against the greyish-green foliage. The wood is quite prickly, the buds are well mossed, and the growth, if allowed, can become quite straggly. This rose is sometimes referred to as 'the Old Velvet Rose', and is found in many old gardens and is dear to many hearts. It is a superb old Moss rose.

'Winchester Cathedral'

Raised by David Austin, United Kingdom
Parentage: Colour sport of 'Mary Rose'
Introduced 1988
Type: English Rose
Height: 2 metres

Hybridisers go through the agonising process of painstakingly hand-pollinating parent plants, waiting for the seed pod to ripen, sowing the seed and hoping enough will germinate for a good selection, and then down the track selecting a plant with the required merits. Nature in this case has circumnavigated the process by presenting the hybridiser with a ready-made rose. This rose is named after the Winchester Cathedral Trust. It is similar to 'Mary Rose' except that it is pure white. It has a lovely scent and the excellent quality of always having some flowers on show.

'Windrush'

Raised by David Austin, United Kingdom
Parentage: English Rose seedling x ('Canterbury' x 'Golden Wings')
Introduced 1984
Type: English Rose
Height: 2 metres

Although not really within the confines of the definition of an English Rose, 'Windrush' nevertheless has a double dose of English in it. It resembles 'Golden Wings' in the colour and size of the flowers, which are quite large (13–15 centimetres wide). The buds are yellow and long pointed, grouped five to seven at a time, and the exquisite blooms open wide and semi-double, exhibiting golden stamens. The flowers are nicely fragrant. This fine rose could quite easily be made to climb.

'Wise Portia'

Raised by David Austin, United Kingdom
Parentage: 'The Knight' x 'Glastonbury'
Introduced 1982
Type: English Rose
Height: 1 metre

The parents of this excellent rose are the same given for 'Wenlock', another David Austin English Rose introduced in 1984. If nothing else, this proves that if someone else were to repeat the process with the same parents it would be practically impossible to achieve the same result. The bush has low, rather spreading growth that is assisted by the large, double, heavy flowers. The blooms come from very fat buds and open wide and well filled with petals of a genuine old rose character. The colour is at first rich purple, paling a little with age. The blooms have a strong old-rose fragrance.

'Yellow Button'

Raised by David Austin, United Kingdom
Parentage: 'Wife of Bath' x 'Chinatown'
Introduced 1975
Type: English Rose
Height: 1 metre

When I first saw this rose in 1985 in England I thought it was the most beautiful thing I had ever seen. This reaction was quite natural when you remember that up until that time I had spent many years looking at and enjoying the classic old roses and their more restricted colour range. 'Yellow Button' has beautiful blooms of the deepest yolk yellow in the centre, becoming paler toward the edges. They are rosette type, quartered, and have a button eye, resembling an old rose in every way, but with a modern colour. A strong fruit fragrance pervades the blossoms and the foliage is a light glossy green. Since its introduction, other roses such as 'Symphony', 'English Garden' and 'The Pilgrim' have joined this colour group.